VOLT
GOODYEAR

LARRY EDSALL

First published in 2010 by Motorbooks, an imprint of
MBI Publishing Company, 400 First Avenue North, Suite 300,
Minneapolis, MN 55401 USA

Motorbooks titles are also available at discounts in bulk quantity for industrial or sales-promotional use. For details write to Special Sales Manager at MBI Publishing Company, 400 First Avenue North, Suite 300, Minneapolis, MN 55401 USA.

To find out more about our books, visit us online at www.motorbooks.com.

Library of Congress Cataloging-in-Publication Data

Edsall, Larry.
Chevrolet Volt : charging into the future / Larry Edsall.
p. cm.
ISBN-13: 978-0-7603-3893-3
1. Volt automobile. I. Title.
TL215.V64E37 2010
629.222'2—dc22

2010025093

About the author:
Larry Edsall was snatched away from a career as a daily newspaper sports editor to become motorsports editor at *AutoWeek* magazine. Before long, he was automotive industry news and motorsports editor and for most of his 12 years served as the magazine's managing editor. While at *AutoWeek*, he drove nearly half a million miles evaluating vehicles on four continents. He left Detroit for Phoenix late in 1999 to help modernize one automotive website, then launched another, iZoom.com. He also writes as a freelance contributor to several automotive and lifestyle publications.

On the frontispiece: *GM*

On the title pages: *GM*

On contents page: *GM*

On the back cover: *GM*

Editor: Chris Endres
Design Manager: Kou Lor
Designer: Cindy Samargia Laun

This edition ISBN-13: 978-0-7603-4381-4

Printed in China

Contents

FOREWORD BY BOB LUTZ

EVER SINCE WE INTRODUCED THE CHEVROLET VOLT CONCEPT CAR AT THE NORTH AMERICAN INTERNATIONAL AUTO SHOW IN DETROIT IN JANUARY 2007, I'VE BEEN SAYING THAT THE VOLT PROGRAM IS THE MOST EXCITING PROJECT I'VE BEEN INVOLVED WITH IN MY ENTIRE CAREER IN THE AUTO BUSINESS. I KNEW IT WOULD BE—EVEN BEFORE THAT DAY—AND I STILL FEEL THAT WAY TODAY, NOW THAT THE VOLT IS READY TO HIT AMERICAN ROADS.

In the aftermath of that introduction, Volt faced a lot of criticism. It was called "vaporware" by some members of the media. Executives from some other manufacturers told the press that the battery technology wouldn't work—many of those same companies eventually climbed aboard the lithium-ion bandwagon. GM was even accused of using the Volt as nothing but a PR stunt to "greenwash" its image. When we introduced the production version of the car in September 2008, Volt critics finally had to accept that the car was real and was headed to showrooms and streets.

The Volt means a lot to GM and to the industry on a variety of levels. First of all, this is solid technology that is going to be proven reliable. It's a practical way that we can electrify the automobile and drastically reduce our dependency on imported petroleum. It's also important to GM to help reinforce and continue its proud history of technological innovation, and to help restore the image of leadership that accompanied that history.

In terms of the impact of the Volt on the automobile industry, I think you'll see lithium-ion technology continue to evolve and be used by the rest of the industry. And as for society in general, I think it can have an enormous benefit. Nearly 80 percent of Americans drive 40 miles a day or less. That means that nearly 80 percent of Americans can commute powered by electricity from the grid, never using a drop of gas.

When we achieve substantial production, and if our competitors do as well and the public takes to this new way of driving—and there's no doubt in my mind they will—we will drastically reduce gasoline and diesel consumption and our dependency on oil. So I think it's very hard to overestimate the importance of the Volt for GM, for the industry, and for automotive transportation as a whole.

With the Volt's appearance on the road now imminent, this book presents an excellent chronicling of exactly how it all happened. GM was extremely transparent in the development of the vehicle, as we promised we would be, and all of those details and more are captured here. For once the old show-biz line, "We hope you enjoy it as much as we enjoyed bringing it to you," isn't a cliché. From where I sit, it's the unvarnished truth.

INTRODUCTION

ON JANUARY 7, 2007, I WAS AMONG HUNDREDS OF AUTOMOTIVE JOURNALISTS WHO GATHERED AT THE GENERAL MOTORS DISPLAY AT THE NORTH AMERICAN INTERNATIONAL AUTO SHOW IN THE DOWNTOWN DETROIT AREA FOR THE UNVEILING OF WHAT WE WERE TOLD WASN'T JUST ANOTHER CONCEPT CAR, BUT A VEHICLE THAT WOULD INTRODUCE US TO GAME-CHANGING AUTOMOTIVE PROPULSION TECHNOLOGY.

When the car's cover was removed, I remember thinking I really didn't care what sort of power source might be inside this compact sedan. I liked the car's design and hoped it would find its way to the road as quickly as possible.

But the big news wasn't the package, it was the powertrain. This stunning vehicle was designed to showcase a technology that General Motors promised would be a significant step in weaning America and its motorists from their dependence on imported oil.

At this point my auto writer's DEW system went to full alert. You remember the DEW line: The Defense Early Warning technology was a Cold War creation designed to alert the United States to a missile attack from the Soviet Union. Well, I'd developed my own alert with regard to technology promises from General Motors.

Early in my automotive reporting career, I'd walked around the Detroit show with a since-retired engineer who for many years had led General Motors' engineering research and vehicle development efforts. An overseas automaker would be unveiling, say, computerized vehicle dynamic control technology or a hybrid powertrain that greatly enhanced fuel economy by linking electric motors and batteries with the usual internal combustion engine.

"We did that years ago," the GM engineering executive would say.

But what good, I wondered, was creating a technology if you were tardy in bringing it to the road-going marketplace?

In ensuing years, I'd hear GM unveil, say, a car with a skateboardlike chassis that supposedly could carry any sort of body configuration—pickup truck by day and luxury sedan by night, for example—or maybe it was the promise that nonpolluting fuel cell powertrains were just around the next corner, "*Really, they'll be here any day now*," and I'd learned to dismiss such whiz-bang claims as auto show promises that would remain unfulfilled.

But when the Volt was unveiled, I heard something I'd never heard before from General Motors executives. They said that while the technological theory was sound, the lithium-ion batteries needed to make the Volt a viable vehicle were yet to be developed.

The concept behind the Volt was not to use an internal combustion engine that constantly consumed imported petroleum nor a hybrid powertrain that could get the car underway on battery power but to actually use a battery pack to do all the driving. That battery pack would be plugged into the electrical grid for overnight charging and would propel the Volt up to 40 miles.

Now remember that when this concept was being unveiled, General Motors was being held up to cinematic scorn for being the company responsible for "killing" the electric car. Except what killed the electric vehicle wasn't General Motors but range anxiety. Drive a traditional car and you might get nervous as you watch the "distance to empty" number on the dashboard display drop lower and lower. Nonetheless, you could take comfort in the fact that, most likely, you would find a gas station within the next few blocks or at the next exit. Not so with an electric vehicle. Even if you found a proper electrical outlet and plug, it would take hours, not minutes, to recharge the batteries.

But this Volt eliminated range anxiety because it came with its own battery recharger—a small internal combustion engine that had no mechanical connection to the car's drive wheels but would run to generate electricity to maintain the battery pack's charge so the car could drive not just tens but hundreds of miles.

At least that was the concept. There was one problem, according to General Motors executives—the lithium-ion batteries needed for such a vehicle did not exist.

Never before had I heard GM executives admit such a thing about their latest and greatest, and there was something about the honesty in their admission that made me think they weren't just blowing smoke across mirrors. There was something in their honesty that told me they really were serious about this concept and its technology. When lithium-ion batteries were ready for automotive application in volume production, so were they.

As it turned out, those batteries were just months away. The Volt not only would become a viable and game-changing vehicle but, in fact, would go from concept to production in record time—despite some amazing obstacles.

While the Volt was in development, General Motors would go through federally mandated bankruptcy and three chief executive officers, and the Volt team itself would see several of its key people leave, though, fortunately, they all left after their vital contributions to the project had been made. Some of the names and faces changed, but the program continued on target.

That program—the quest for a car and technology that might, at last, make a significant step toward American automotive energy independence—is the subject I hope you'll find revealed in the pages that follow.

CHAPTER 1

THE SPARKS START TO FLY

You might have thought that Bob Lutz would take a while to bask in the warm and glorious afterglow of the 2006 North American International Auto Show in downtown Detroit.

As vice chairman and chief of global product development for General Motors, Lutz was in the driver's seat, his hands on the steering wheel, as he maneuvered the Chevrolet Camaro concept car through Cobo Hall to its official unveiling the morning of January 9. A muscular and sleek silver missile, the Camaro concept car received overwhelming acclaim, not only as a vehicle that would reignite the fun factor that had made cars so vital for the baby boomer generation, but as one that could reinvigorate General Motors itself as it tried to steer its way back to profitability despite an increasingly competitive marketplace.

But Bob Lutz didn't get where he was by sitting back and basking. A former U.S. Marine fighter pilot who continues to fly jets and helicopters well into his 70s, Lutz also had a jet-propelled automotive career, taking him from General Motors of Europe to BMW to Ford to Chrysler and finally back to General Motors—and onto the cover of the book, *Six Men Who Built the Modern Auto Industry*, where Lutz was posed as part of a second-half-of-the-twentieth-century automotive Mount Rushmore, alongside Henry Ford II, Soichiro Honda, Lee Iacocca, Ferdinand Piech, and Eberhard von Kuenheim.

Lutz had made BMWs accepted—and desired—as the ultimate driving machines. He had helped make the minivan an American family favorite. He had championed high-performance vehicles, such as the Viper at Dodge and the Camaro at Chevrolet. Though neither a designer nor an engineer (his degrees from the University of California–Berkeley are in business), his personal automotive passions strongly influenced the look and feel of vehicles everywhere he worked.

In January 2006, Bob Lutz' automotive passions were stewing to a boil.

"I was smarting very much under this tidal wave of favorable publicity toward Toyota as the wonderful car company that can do no wrong," Lutz remembers.

The perception—in the media and held by the car-buying public—was that Toyota, with its fuel-sipping Prius, a gasoline/electric hybrid that was leading the way green, was, as Lutz put it, "environmentally conscious, obviously not driven by the profit motive at all but by the intense desire to do good for mankind."

At the same time, Lutz knew that "[GM] was increasingly seen as a laggard in technology," not only slow to get into the hybrid vehicle game, but about to be featured in a most unflattering role in a documentary movie for pulling the plug on the EV1, the plug-in electric vehicle GM had launched in the late 1990s.

Lutz asked himself: If he were Toyota, what would he do next to reinforce and even accelerate this environmentally friendly, technologically advanced image? And, he wondered, what might he do to demonstrate General Motors' extensive (if perhaps unseen by the car-buying public) technological capabilities? (Lutz contends that GM engineers and scientists had developed gasoline/electric hybrid technology much like that used in the Prius, but GM chose not to put the system into series production at the time.)

Rather than bask in the praise surrounding the rebirth of an icon, the Chevrolet Camaro, what Lutz decided to do was to champion the development of what he knew would, or at least ultimately could, be an even more significant vehicle—an electrically propelled, General Motors–designed and built concept car to unveil at the next Detroit auto show, in January of 2007.

"All of a sudden," Lutz says, "Tesla announced that they've got this two-hundred-mile-range electric car with six thousand eight hundred and thirty-one laptop batteries all wired together, and it's going to cost a hundred thousand bucks, and it's going to do zero-to-sixty in four seconds, and it's going to have one hundred forty miles per hour top speed and . . . "

Tesla Motors was a Silicon Valley startup funded by Elon Musk, who used his degrees in physics and business to launch Internet companies including PayPal. Musk wanted a non-polluting yet high-performance sports car to drive. When he couldn't find one, he helped electric propulsion systems engineer J. B. Straubel and others launch Tesla Motors to build them.

Facing page: The Chevrolet Volt, General Motors' revolutionary electric vehicle with extended-range capability. *GM*

Below: Bob Lutz was the center of attention when Chevrolet unveiled the Camaro, its concept for a modern muscle car, at the Detroit auto show in 2006. *GM*

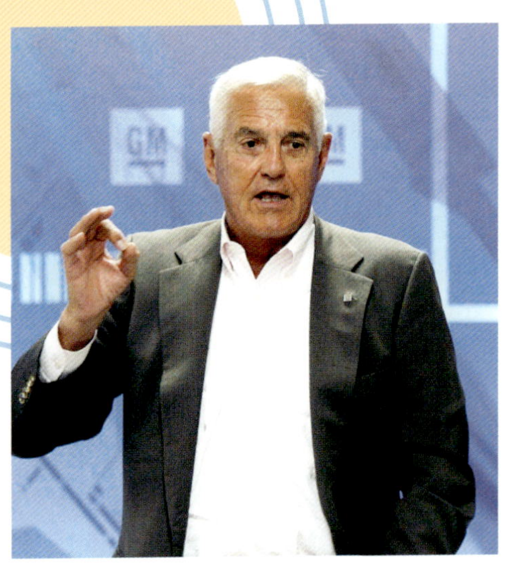

Despite the buzz created by Tesla's announcement and Lutz' personal enthusiasm for an electric-powered vehicle, Lutz' idea to develop a battery-powered concept car was not eagerly endorsed by other high-level General Motors executives.

"I talked about it in a couple of meetings at GM, and it was always met with hoots of derision," Lutz recalls.

In fact, he says, then-GM-chairman Rick Wagoner dismissed the idea, saying that GM had lost a billion dollars on the EV1 and now Lutz proposed to lose another billion should his electric car concept be approved for production.

Wagoner wasn't alone in his opposition. Those involved in GM's fuel cell development program—which included a demonstration fleet of some 100 Chevrolet Equinox vehicles powered by electricity-producing hydrogen fuel cells—saw no reason to do a more conventional battery-powered vehicle in the meantime. They preferred to wait until fuel cells were ready for public consumption.

"At which point I said we can't wait until 2025 to do something," Lutz countered matter-of-factly.

After leaving Chrysler and before returning to General Motors in 2001, Lutz had been chairman and chief executive of automotive battery supplier Exide Technologies for three years.

"I knew that lithium-ion [battery] technology was certainly on the cusp of being ready," Lutz explains.

But when he suggested that to his fellow GM executives, "I was pooh-poohed. Everyone said lithium-ion won't work. 'Let us have you talk to our advanced battery people, Bob, and they will make you understand why lithium-ion won't work for cars.'"

Admitting that he was not an engineer, Lutz agreed to listen to the company's experts, "and battery expert after battery expert came in and said the problem with lithium-ion is that they're energy batteries, they're not power batteries, and for a car you need a power battery.

"And I said what about the batteries used in power tools? Well, they said, they're especially designed for power

Even into his 70s, Bob Lutz was flying jet fighters. Lutz had championed high performance throughout his automotive career, but now he was pushing, and pushing hard, for General Motors to create not another fuel-burning hot rod but an environmentally friendly, electric car. *Peter Yates*

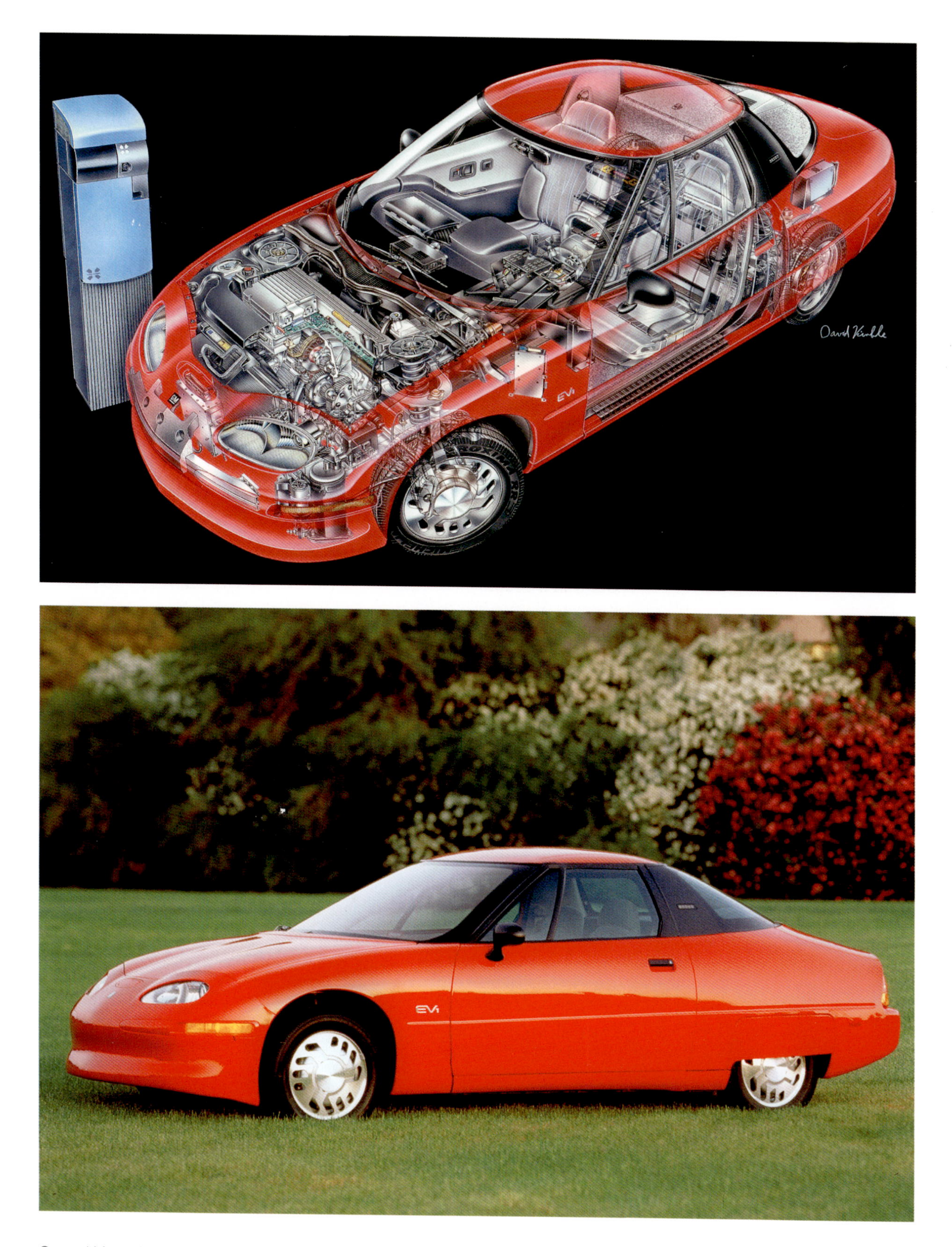

General Motors' most recent experience with electric vehicles was the EV1, a small two-seater with limited driving range and thus limited appeal. When GM withdrew the car, it was accused of killing electric cars forever. *GM*

The difference between Lauckner's electric vehicle and the typical gasoline/ electric hybrid was that there would be no mechanical connection between the gasoline engine/ generator and the vehicle's drive wheels.

tools and they wouldn't cool very well in an automotive application . . . and et cetera . . . and et cetera . . . and I kept getting these stories."

So, Lutz countered, "How is it that all of our experts tell me that we can't do a lithium-ion-powered car and here's this little outfit in California that does one, admittedly expensive, [but] I didn't know at the time it was going to take them almost four years to get it into the market."

Nonetheless, Lutz was determined.

"At this point," he recalls, "I said, 'I think as head of product development, I'm really going to talk to my guys about a fully electric, lithium-ion prototype for a show car.'"

"Well, all right," Lutz remembers being told in an executive meeting, "but come back here before you do anything."

And that, Lutz mentions modestly, was his personal contribution to what would become the Chevrolet Volt.

"I got the idea of a lithium-ion-powered electric car at least to the point where they didn't throw rocks at me at the meetings."

Actually, Lutz went well beyond that point. He talked to his guys about it, and to one guy in particular.

"I went to talk to Jon Lauckner," Lutz recalls. "Jon is an engineer and a brilliant schemer."

Jonathan Lauckner had joined General Motors in 1984, working in powertrain and vehicle engineering before moving into product planning. In the 1990s, he directed GM market and business analysis for the Asia-Pacific and Latin American regions and became director of export operations and trade development and director of planning for GM do Brasil. At the turn of the century, he moved to Europe, overseeing development of compact vehicles for Opel and Vauxhall. Four years later, he became global vehicle line executive for GM's Epsilon compact vehicle architecture. He returned to the United States in 2005 to become GM vice president for global (vehicle development) program management.

Like other GM engineering experts, Lauckner also told Lutz what he didn't necessarily want to hear.

"All-electric is not smart," Lutz recalls Lauckner's words, "because no matter how much [of an] expensive battery you put in it, the range is always going to be limited, and if you use one hundred percent of the battery—full charge to full discharge—the battery life is going to be seriously jeopardized."

However, instead of dismissing Lutz' idea, Lauckner became a co-conspirator.

Lauckner proposed doing the concept not as a purely battery-powered vehicle but with what Lutz remembers Lauckner referring to as a "smallish" lithium-ion battery pack, "call it sixteen kilowatts," Lutz repeats Lauckner's words, "so we charge it to maybe eighty percent and discharge it to something like thirty percent, to where the battery isn't even breathing hard, and that way we can be sure of a ten-year life, and what we do is basically a series hybrid: We'll have a small gasoline engine driving a generator, which will feed the battery once the initial charge runs out."

The difference between Lauckner's electric vehicle and the typical gasoline/electric hybrid was that there would be no mechanical connection between the gasoline engine/generator and the vehicle's drive wheels. In most hybrid applications, sometimes the electric motors drive the wheels and sometimes the gasoline engine does. But the engine in Lauckner's concept would only generate electricity to feed the electric motors that actually turned the wheels. The Lutz/Lauckner concept would be a true electric-driven vehicle.

Lutz resumes the story: "He took out a piece of paper and his expensive fountain pen, with which he drives me nuts—he always pushes against the nib and the nib vibrates and little sprays of ink come out and half the time the nib goes through the paper—but he did the calculations very quickly.

"He says, 'OK, lithium-ion . . . smallish car . . . it's about five miles per kilowatt hour of energy, and if we use eight kilowatt hours, that would be forty miles, which is probably a good range for most people, and that way we could keep a reasonably sized battery pack. . . .'"

"I started doodling some of the things that we ought to take into account," Lauckner says. "First of all, I told him that to do a pure battery electric was possible, but the problem was [that] to get enough battery on board to give it a decent range, the battery pack would have to be enormous—very, very large—because the type of vehicle we talked about was not a two-seat roadster. It was a vehicle that was more of a family-sized vehicle, something that would be around a compact or midsize vehicle, which means it would have to have far more battery than the Tesla Roadster, which is an all-aluminum, two-seat, very tight sports car.

"My comeback to Bob was: 'Why don't we just put enough battery on-board so the vast majority of people would drive electrically, let's say seventy to eighty percent of the people would be able to drive electrically in their daily commute, and then we'll put a small internal combustion engine and a generator to generate more electricity when the battery is depleted. That way we'll have a vehicle that can have a range of hundreds of miles, the way most people expect if it's the first car in the family, not just a commuter car.'"

A bonus, Lauckner realized, is that "we take the whole discussion of infrastructure off the table, because, as we learned in the EV1 days, if you're talking battery electric, you can't talk about battery electrics unless you're going to talk about infrastructure—public infrastructure—to charge them."

Lauckner's vision was that people would recharge this new GM electric car overnight by plugging into an electric outlet at home.

Further, he said, creating an electric vehicle with an onboard gasoline generator, a range-extender if you will, would be much more cost-effective to create than trying to package an enormous battery system to provide the same hundreds of miles of range.

"I wanted to go with a much smaller battery, something we could put in the tunnel [that runs between the front seats and typically houses a driveshaft or other underbody components]," said Lutz. But such a small battery might provide only around 10 miles of pure-electric propulsion.

Before their initial conversation was over, Lutz recalls Lauckner had "figured out how many kilowatts the engine would have to produce, how many kilowatt hours, the basic packaging layout, what the lithium-ion battery would look like, the cell size."

"Jon convinced me that fully electric makes no sense because to get a one-hundred-mile range, which is kind of a minimum you need, requires a twenty-kilowatt-hour battery, which would be enormous, and you're cramming the whole vehicle full of battery," Lutz continues. "And you're using one hundred percent of the battery. And on a cold day the pure electrics won't start in North Dakota in the middle of winter, so its mainly a California and southern states play, whereas this concept will, of course, light off on its gasoline engine, and the gasoline engine heats the battery; after five or six minutes the computer will shut down the gasoline engine and it will take over electrically. The pure electric can't do any of that.

"It was very easy to convince me that forty miles of electric range, followed by another two hundred and fifty or three hundred miles of gasoline range, gave the customer the maximum security of ensuring mobility under any and all circumstances while permitting the

Jon Lauckner convinced Bob Lutz that to make his electric car concept practical, it needed an engine generator to create electricity on-board to extend the vehicle's range. *GM*

benefit of eighty percent of American daily driving needs to be accomplished purely electrically.

"This was a very compelling notion." Lauckner recalls that his conversation with Lutz took place in early January, while the auto show was still underway, and that things progressed very quickly afterward.

"I knew a guy named Greg Adams," says Lauckner. "He was in charge of business development for GE Plastics. We knew each other from the days we were in Europe. We did a very innovative roof module system for the Opel Zafira that basically used a lot of plastics, and although we didn't try to do it, it made a big splash in most of the trade publications on plastics. We maintained contact, and when I came back to the United States from Europe in mid-2005, he got in touch with me and said we need to do some innovative things here."

On January 17, Lauckner set up a dinner that had Lauckner, Lutz, Adams, and some of Adams' associates from GE Plastics at the table.

"We talked about getting together, GM and GE, to create the concept car," Lauckner recalls. GM would develop a new propulsion system and a stunning design, and GE would supply various advanced plastics and other components to be showcased.

Almost immediately, Lauckner put together an internal working group at General Motors. "It would have been myself, Larry Burns [the head of research and development and the company's fuel cell champion], Bob Purcell [advanced technologies], Chris Preuss [from public relations], Jack Kiebler [from product development], Anne Asensio [from advanced design], Jon Bereisa [advanced engineering], and maybe a couple of other people," he says.

"The idea was to do a unique and profound concept car that would give us a very leading edge position in advanced technology and design and likely preempt some of the other competitors that would show these types of vehicles."

At this point, the concept was being referred to as the iCar, "because we wanted to be as ubiquitous and profound as the iPod, which was still very new at the time," Lauckner remembers.

And yet, not everyone in the conference room that day at GM's Vehicle Engineering Center in Warren, Michigan, was thrilled with the concept, nor were they alone.

"The mechanical hybrid guys in the company all surfaced and said, 'We don't like this,'" Lutz remembers. "'This is not an efficient way to convert energy.' I said: 'What do you mean?' They said: 'Well, series hybrids are notoriously inefficient because of the losses of moving mechanical energy to electrical energy and then back to mechanical energy, whereas a parallel hybrid, whichever energy source is the most efficient, like gasoline, cuts in when gasoline is the most efficient [and] electric cuts in when electric is the most efficient. . . . If we do this concept at all, may we respectively suggest that we run it to where it will be ten minutes on the internal combustion engine to keep the batteries topped up and maybe ten

Bob Lutz again is the center of attention, this time at the 2007 North American International Auto Show, where General Motors unveiled the Chevrolet Volt concept car. *GM*

The concept car version of the Chevrolet Volt was unveiled in Detroit on January 7, 2007, less than a year after Bob Lutz started promoting the idea within General Motors. *GM*

to fifteen miles on battery and then ten miles on internal combustion and that way we'll get the most range.'

"I said: 'Look, guys, you don't fully understand. This is not about maximizing range or computationally getting the maximum efficiency. This is about permitting the average American to drive fully electrically most of the time but with the backup of a gasoline engine so that if I'm on a local errand and I get a cell phone call saying that my mother is in bad shape in Chicago, I don't have to go home and get my other car. I just turn around and head for Chicago, knowing that I'm going to get there and knowing that I'm going to get there at seventy and eighty miles per hour if need be. That's what it is about.'

"It's not about trickily trying to get the best EPA number or the farthest range or being able to say overall it consumes fewer BTUs than any other vehicle on the planet. That's not what it's about. It's about electric drive with the assurance of gasoline backup."

Lutz recalls there were several rounds of such discussions with "the learned engineers and scientists trying to dissuade me, but every time I had Jon Lauckner with me and, of course, he being an engineer, which I'm not, he was able to head them off at the pass every time."

In September 2008, Bob Lutz drove the production version of the Chevrolet Volt onto the stage for its official introduction as part of General Motors' centennial celebration. *GM*

CHAPTER 2

The Concept Takes Shape

By February 2006, Bob Boniface had learned that Bob Lutz wanted to do a concept car for the 2007 Detroit auto show that would have an even larger impact than the spectacular Chevrolet Camaro concept that had been revealed just a few weeks earlier at the 2006 Detroit show.

Any such vehicle would have to be stunning to the point of game-changing.

"The Camaro reveal was enormous!" says Boniface, who remembers he was still caught up in what he describes as the "usual hangover" after the Detroit show, a period when the euphoria finally starts to wear off after Detroit's annual turn in the global automotive spotlight, unveiling its latest cars and concepts and hosting auto industry leaders and the reporters, photographers, and broadcasters who proclaim the news from the automotive equivalent of the Academy Awards or Fashion Week.

As head of what is known within General Motors as Design-North, Boniface and his design team were involved early on in all of the company's concept vehicles.

Design-North is GM's home-office advanced design studio. While located within architect Eero Saarinen's mile-square masterpiece, the General Motors Technical Center, where production vehicles are styled and shaped, schemed and sculpted—and engineered—Design-North takes its name from being hidden away at the far north end of the campus, where its work can be kept secret, not only from outsiders who might be attracted to the big styling dome on the south end of the campus, but from GM staffers who don't have need-to-know access to the company's possible future projects.

At first, Boniface recalls, his instructions on Lutz' latest concept were not very clear, "so the guys started sketching the traditional cars without wheels and things that were not as automotive as perhaps they should be.

"Upon showing the sketches to Bob, he said, 'No. No. What I want is an electric vehicle.' That gave us clarity. Now we knew we were working with a vehicle that actually had four wheels and people actually sat in and you drove it, you didn't fly it," says Boniface.

The designs for the concept car were coming into focus because the concept for that car was gaining its own clarity.

Jon Lauckner, Lutz' champion for the project and the head of General Motors global vehicle program development, was continuing to meet with his small working group, defining the initial iCar concept. "The scope and mission" is how Lauckner terms the group's work to verify the concept's viability and feasibility. The working group reached its conclusion before the end of March.

Immediately, Lauckner established a small project team of people with the technical expertise to refine the concept, to nail down the basic parameters.

"We knew it would be a compact or midsize car," Lauckner says. "We knew it had to have four- or five-passenger seating. We knew it had to have an engine/generator that would generate a certain amount of electricity because, quite frankly, it takes a certain amount of power to drive a certain mass down the road.

"But you need to get specific. You need to run simulations, to make sure, in fact, that what you think is the case is actually validated by simulation because you wouldn't want to go out there and introduce a brand new concept and have everybody say, 'Well, it's a nice idea, but they made some very flawed assumptions and this thing absolutely won't hold water.'"

On April 28, Lutz, General Motors' global design director Ed Welburn, and advanced design chief Anne Asensio reviewed designers' proposals for the concept car. Exterior designs were presented in sketches and scale models—five from the advanced studio in Warren, Michigan; two from GM's advanced studio in England; and two from the company's advanced studio in North Hollywood, California.

One design already had emerged quickly as the favorite within Boniface's Design-North studio. As Lutz and the company's top design managers winnowed the entries, it was clear why.

"Jelani had just captured the grace and proportion we wanted for the concept car, and we fell in love with it," Boniface remembers.

Jelani Aliyu, a native of Nigeria, joined the GM design staff in the summer of 1994.

Aliyu grew up in northern Nigeria, just south of the Sahara desert. "I've always been fascinated by cars," he says,

Above: In mid-July 2006, Bob Boniface, design director in what at the time was called the iCar Systems Studio, closely checks the full-scale clay model of what will be unveiled as the Chevrolet Volt concept car. *GM*

Opposite: Bob Lutz, General Motors vice chairman of global product development, introduces the Chevrolet Volt concept car to automotive journalists from around the world at the 2007 North American International Auto Show in downtown Detroit. *GM*

He often went with what we recommended. That was fun."

Aliyu's father was an educator. He had been schooled in London and taught there before joining the Nigerian foreign service and working in Sudan. He returned home to work in the ministry of education. "When the movement began to go back to a democracy [after nearly two decades of dictatorial military rule], he was one of the six national commissioners [who] prepped the country and got it ready for elections," Aliyu says.

Aliyu remembers that he always wanted to be a car designer. "I loved drawing. I drew people and houses and a lot of cars. People would say, 'No, why would you want to be a car designer? You couldn't get a job here.' But that was what I wanted to do and I was determined to do that."

To launch his dream, Aliyu spent two years in architecture school in Nigeria while getting ready to pursue his car design education and career overseas.

What he really wanted to do was to go to school in Italy, home of those gorgeous Ferraris and Lamborghinis. "But I learned that to go to school there you had to learn Italian," he says. "It is a beautiful language, but I didn't want to have to learn it. I wanted to get straight into car design."

He considered going to England to study, though at the time the Royal College of Art offered only a graduate-level program in transportation design.

Top: Bob Boniface (left), Anne Asensio (center), and Young-Sun Kim examine details of the Chevrolet Volt just weeks before the car is finished and ready to unveil at the 2007 North American International Auto Show in downtown Detroit. *GM*

Middle: As the Chevrolet Volt concept car is nearing completion, design manager for advanced vehicles Young-Sun Kim (left), E-flex creative designer Jelani Aliyu (center), and Anne Asensio, executive director for advanced vehicle design, consider the details of a prototype for the car's grille. *GM*

Bottom: Max Larroquette, engineer on the Volt concept car project, seems pleased with the way the proposed grille fits on the front of the car. *GM*

adding that while there were no sports cars in Sokoto, he and his brothers and friends "fell in love" with every photo they saw of a Ferrari or Lamborghini.

"My brothers and my dad were very much into cars," he added. "When we were young and my dad would buy a new car, he would narrow it down to three and ask us which one we wanted.

Volt concept exterior designer Jelani Aliyu says he was inspired by lions on the Serengeti in his African homeland when he started sketching his ideas for the car. *GM*

A cousin loaned Aliyu a book that listed all American colleges. "I went to the art and design section, and I picked out a couple with names that sounded good," Aliyu says. The two that sounded best to him were the Art Center College of Design, a famed automotive design school in Pasadena, California, and the College for Creative Studies in Detroit.

"I decided to go with CCS because of its proximity to the Big Three," says Aliyu, "and they also offered dormitories, and coming from out of the country, that would be much easier for me."

Aliyu put together a portfolio of his designs—primarily automotive, but he also included a pair of sneakers to demonstrate industrial design—and sent it to Detroit. "I got admitted," he says. He also applied for and received a scholarship for overseas study from the Nigerian government and arrived in Detroit in January 1990.

After graduation, Aliyu was dealing with the complications of obtaining a visa to work in the United States and actually was getting ready to go back home because his student visa was expiring. He was using a pay phone on the ground floor of his apartment building to call the Nigerian embassy and arrange for an airline ticket home when, through a window, he saw a friend and fellow graduate washing his car. When Aliyu couldn't get through to the person he needed at the embassy, he went outside.

"Where have you been?" the friend asked. "General Motors has been trying to get a hold of you."

Instead of the Nigerian embassy, Aliyu called GM Design. Yes, he was told, they were eager for him to come in for an interview.

"I almost missed it and went back home," Aliyu says.

Not only was he hired by GM, but he was assigned to work with Ed Welburn, who at that time was in charge of a portfolio studio, an advanced design studio where Aliyu worked on proposals for future Pontiacs and on concept vehicles.

Next, Aliyu worked on the exterior design of the Oldsmobile Bravada sport utility vehicle. After that, he worked very briefly in another advanced studio before being recruited to join Liz Wetzel and the team designing the Buick Rendezvous. Aliyu became the lead interior designer for the crossover utility vehicle.

His next assignment was overseas, in Europe, to work on the Opel Astra and other projects. "That was quite an experience," he recalls, not only because of the work but because of living for more than a year in Germany with its infamous autobahns and sports cars.

Aliyu returned to Warren and became the lead exterior designer on the Pontiac G6 sedan, was part of a group doing exterior design proposals for future GM compact cars, did the interior for the full-size Cadillac DTS, and then reported to Design-North for his next assignments, working on a couple of projects before Boniface called a meeting to talk about a show car Bob Lutz wanted to do.

"I thought, 'Wow! This is a great opportunity,'" Aliyu says. "It was in tune with my personal philosophy. I really believe that we should strive to do things that haven't been done before.

"A couple of years earlier, we had what they called the 'diagonal slice' meeting where they invite people from all levels to meet with higher design executives." At that meeting, Aliyu read from a one-page philosophical statement he had written, which included something close to the following:

Humanity is on the verge of a major philosophical evolution, an evolution that will dramatically and forever change the very pattern of the human experience. This will be brought about by exponential advances in the sciences, art, and technology. A significant result of this new way of thinking will be our relationship with our environment.

"I thought that nature would no longer be seen as a force to conquer, but rather it would be seen as a medium from which to learn, to draw inspiration, and to live in harmony," Aliyu adds. "I proposed that General Motors should really strive to do things that when they came out, people would say, 'That's impossible!' When this program came on, I saw this as my chance to do what I've always believed in."

Aliyu says he drew his inspiration for his concept design from his experience growing up in Nigeria—and from staying attuned to his native environment even while living in the United States.

The Volt concept car emerged from sketches by Jelani Aliyu, a GM designer from Nigeria. *GM*

"In Nigeria, you get away from the cities and in a couple of minutes you're out in the wild," he explains. "During my stay at CCS and here at GM, I spent a lot of time watching the Discovery Channel, seeing the animals and undersea life.

"We live on a truly magical planet. It's a gigantic orb of life hurtling through space. Everywhere we look around us, we're surrounded by the wonder of our natural world, from the tiny leaf, only a fraction of a millimeter thick and yet a highly efficient factory, to the amazing sea rays that glide within the deep waters of the Atlantic Ocean. Planet Earth is a perfect balance of beauty and practicality.

"So I drew from that as inspiration for the Volt, something that is really a perfect balance of beauty and practicality—sleek, low-profile, well-balanced, well-proportioned. Dynamic, and yet practical and doable."

A lion on the Serengeti was what he had in mind. In automotive terms, he says, that would be a sports car but with four doors.

"I was watching the Discovery Channel one night, and I was just doing tiny doodles, a couple of inches long, and I did a lot of them," Aliyu recalls. "I brought them into work the next day, scanned a couple of them in, and brought them onto my computer."

He then "sketched over" the doodles, working them into a complete vehicle, and added a background, a tropical island setting with palm trees and blue water, to the horizon.

"When I put that up [on the board], everybody said, 'That's really good.' That was really the beginning of the Volt in terms of the way it looks." But the way the concept car would look was just one, albeit very significant, piece of the puzzle.

Lions hunt at night, and the Volt concept takes on a dramatic look after the sun sets. *GM*

Front three-quarters, full face, profile, and full rear images reveal the details of the Volt concept car's design. *GM*

Overhead shots show the details of the Volt's see-through roof, which, like the fixed side windows, was produced in conjunction with General Electric's plastics division. Lightweight GE materials also were used in the hood, door panels, rear deck lid, front fenders, steering wheel, instrument panel, and air bag chute. *GM*

Jon Lauckner had set up a small project team of engineers to work through the various powertrain alternatives and practicalities, to study alternatives and their viabilities for development.

"Remember, at the time we were doing a concept car for an auto show. We weren't doing a full-scale production design," Lauckner says.

Nonetheless, by early April, he had Tony Posawatz leading a group that included Nick Zielinski, John Bereisa, and a few others selected because of their specific experience and expertise.

"They were working through the various ideas of what the car would be, how big it would be, its wheelbase and track, and so on and so forth—the natural things you have to do to take an idea from a rough definition into something that you can explain to other people and you can begin to physically represent," Lauckner explains.

One stream of activity for the team was to define the concept's propulsion system, which couldn't be simply a pie-in-the-sky science fair project, Lauckner says. It had to be something that would be credible—to the point that it actually might become the basis of a propulsion source not just for a one-off, hand-built concept vehicle but perhaps for future General Motors factory-produced production vehicles.

Much of the team's work had to be done quickly. Of course, the final design, especially of the car's interior, couldn't really get going until specific dimensional requirements were established, and those couldn't be done until at least the major components were defined. Because seemingly every piece of a concept car is unique and must be designed, fabricated, and hand-assembled, the concept's construction usually demands around 40 weeks. But it's one thing to do a concept car when you can plug in an off-the-shelf powertrain; it's a much more daunting challenge when you're not only designing a vehicle but a revolutionary propulsion system as well.

Here it was, the spring of 2006, and the concept was scheduled to be completed before the Christmas holidays, by December 22.

Leading that engineering and project team would be Tony Posawatz, who admits he had to be convinced to accept the assignment.

"I'm really not interested in a concept car. I do *real* cars—tens and hundreds of thousands of real cars," said Posawatz, who was the most experienced GM vehicle line director, which is an engineering management position that involves such things as making sure a vehicle's development program is kept on time and within its budget while also leading the various cross-functional efforts. Posawatz not only had experience leading small but efficient engineering teams but also a reputation for getting things done without a lot of bureaucratic structure or budgetary support. He could take projects to completion and was not averse to taking risks. These were talents GM needed on this new project.

At first, Posawatz was the only engineer assigned to the program on a full-time basis. "Everyone else was a borrowed resource," he explains. "Each had some unique and special skills and background in the idea we were pursuing, both in developing a concept car and in parallel trying to figure out if this was technically feasible and had a commercial future.

"Nick and John were the primary guys. Nick was to develop the vehicle architecture, and John, who was the chief propulsion engineer for the EV1, was to develop the propulsion system architecture."

Nick Zielinski had been chief engineer for the Chevrolet Equinox fuel cell vehicle and the Sequel fuel cell concept vehicle. He also worked on GM's hybrid programs. He served as chief engineer for the iCar/Volt concept, with Rich Lannen joining his group as program manager and liasion with the design staff on packaging the components with the concept car.

Posawatz had been a GM scholar as an undergraduate. He started with the company as a production foreman, later working in quality engineering in GM manufacturing plants before being sent for an Ivy League MBA under the GM Fellows program. When he returned, he worked on the corporate financial staff, eventually becoming planning director and vehicle line director for full-size trucks. After six years, during which he helped develop GM's QuadraSteer technology, the Chevrolet Avalanche, and the Cadillac Escalade, he worked on hybrids and then became the vehicle line director for premium mid-luxury cars—from the Cadillac DTS to the Chevrolet Impala.

From his perspective, the typical one-off concept car wasn't very interesting. But this iCar project would be different.

"I was attracted to innovation and challenges," says Posawatz, who found this concept to be, well, not merely interesting but also intriguing.

One particularly intriguing aspect was the idea of developing not simply another hybrid but an electric vehicle that would have an extended range—a range beyond that provided by plug-in battery power, that would use a small

Below and page 28–30: The Volt concept's interior was designed to showcase "near-term" technologies and the innovative use of materials and to take advantage of the ingenious amount of ambient lighting provided by the car's clear roof. *GM*

gasoline- or ethanol-fueled engine or small diesel engine or even a fuel cell stack to generate electricity to recharge the battery pack, thus eliminating the range anxiety inherent in more traditional electric vehicles.

Posawatz didn't see this as the creation of just another concept car, but of a vehicle architecture that would provide the foundation for all sorts of future powertrain development—a bridge toward fulfilling the promise that had been presented several years earlier when General Motors showcased its Autonomy concept car, a revolutionary vehicle with all propulsion and mechanical systems contained within a skateboard-style chassis. The idea was that any of a variety of body components could be attached to the skateboard, making it a platform for everything from a high-performance sports car to a family sedan, crossover utility vehicle, or pickup truck.

The challenge for Posawatz and his small team was to take the discussion from "back of the envelope and napkins' chicken-scratching" to a viable plan for a vehicle that could travel 40 miles after being plugged overnight into an electrical outlet and then continue to roll hundreds of miles farther with an on-board generator providing fresh power to the battery pack.

"Making a pure electric vehicle would have been a much easier technological challenge," Posawatz says. But, he adds, developing what came to be called an E-flex—flexible electric—system would enable all sorts of future development opportunities, and that, he adds, "is a big deal."

"Making a pure electric vehicle would have been a much easier technological challenge."

—TONY POSAWATZ

At first, the team sorted ideas based on common sense and basic engineering judgments. Soon, they were using math models and simulation, though "we didn't have a lick of hardware for quite some time," Posawatz recalls.

"In the middle of 2006, we had done enough preliminary analysis where we began exposing the idea to the rest of the company," Posawatz says. "We said, 'OK, here's an idea; there's what our numbers tell us we can do.' So in the summer we started shopping it around to the technical experts, and we got our fair share of slings and arrows sent our way. But we also got good feedback. By the fall, we had enough people aligned, though not everyone was on board, but we'd handled the technical shots to the bow. We could do this."

And, he adds, perhaps only General Motors could have done such a vehicle because of its experience with the EV1 and because it had done so much work with fuel cell vehicles and with hybrids and had so much experience with alternate and biofuels. "No other company had the experience we had," he says.

"Early in the program we were really just working with the concept and trying to come up with the right size and collection of components to do a credible vehicle," adds Zielinski. "We were trying to have a concept with a lot of strong analysis behind it that we could transition to a real program if there was enough interest."

At first, however, Zielinski found very little interest in the project within some parts of GM. For example, when doing a hybrid version of a vehicle, the powertrain engineering organization was responsible for the hybrid propulsion system while engineers, such as Zielinksi, would integrate that system into the specific vehicle. But "when we came forward with this concept, it wasn't strongly supported by our powertrain activity," he notes. "We were pretty much on our own."

However, he says, once the project was ready for peer review, "powertrain did support validating our modeling and said our integration approach was valid."

Jon Lauckner reviews the progress of the concept car at the time:

"We had this small team and they're working away. Design's going. Guys are working on the propulsion system.

"We have a process inside General Motors called the peer review process where basically somebody who has a technical idea or concept has a peer review with other knowledgeable colleagues where they look at it very carefully, ask questions, [and] make comments and observations. The whole idea of the peer review process is to make sure you have the most solid technical concept available, that you've thought through any of the pitfalls.

"In the third quarter of 2006, there were a lot of GM peer reviews of the technical feasibility of the iCar propulsion system concept, overall as well as some of the propulsion system performance metrics."

Lauckner remembers Norm Bucknor playing a vital role in the simulation studies. "We'd go back and forth on e-mail about what Norm had simulated and what it meant and on future simulations we needed to do."

"At the end of August, we had a GM/GE review," Lauckner adds, "GE had been working with design, and we confirmed the direction of the concept and how were going to collaborate."

To reduce mass—and thus enhance potential fuel economy—the concept would have doors and a hood made from GE's Xenoy iQ high-performance thermoplastic composite (HPPC); a see-through roof, rear deck lid, and fixed side glazing made with Lexan GLX resins and Exatec coating technology; front fenders made with Noryl GTX resins; a steering wheel and instrument panel with integrated air bag chute made with Lexan EXL resins; wire coating with Flexible Noryl resins; and Xenoy iQ resins in its energy-absorbing systems.

Yes, lithium-ion batteries would be used in the concept, but as GM would admit at the concept's unveiling, the actual batteries needed to take such a vehicle into production did not yet exist.

"We were looking at two types of cells," Lauckner explains. "One type is what we call prismatic. We also were looking at cylindrical cells. We created two battery packs for the concept car, one showed cylindrical cells and the other showed prismatic cells." It wasn't until just days before the show that the decision was made to unveil the car with the cylindrical cells in place.

"We looked very deeply into the packaging of the engine," Lauckner says. "We started out with a fairly large engine—a one-point-eight liter, naturally aspirated, and I could never get my head around the fact that it was such a large engine for what we wanted to do with it, to use it as a stationary generator, or what we now call a range extender. For a lot of driving, you wouldn't be using the full displacement of the engine.

"I remember in the fall of that year I called a guy in powertrain who I knew from my days in Europe—Uwe Grebe. I said, 'This thing just doesn't feel right to me, feels like the engine is way too big. For a concept car I think we ought to take a look at a different engine altogether.' What we settled on was a one-liter, three-cylinder turbocharged engine, an engine we were using in Europe in a naturally aspirated version. But in order to get to our target power output when the range extender needed to work really hard, at full throttle or at part throttle up a grade, we needed more output, so we needed a turbocharger to give it additional power. That happened in the September–October time frame."

Another late change was the decision to reduce the size of the concept car's fuel tank. Based on the team's calculation, the original tank would have provided more than 600 miles of range, more than was needed in a vehicle that many people would drive less than 40 miles a day.

On October 25, 2006, Lauckner presented the iCar and E-flex concepts to General Motors' automotive product board. The presentation included the global potential for the platform and for the powertrain, including the possibility of making it the basis not only for the proposed plug-in with on-board range-extending generator but for a pure battery-electric car and a fuel-cell-powered-vehicle.

It was around the time of the meeting that the name Volt was selected for the concept and E-flex for the new propulsion architecture. The other name considered seriously for the concept car was Electra, Greek for "brilliant" and the name of the top-of-the-line Buick model from 1959 to 1990. GM decided that a fresh new name was needed for this new car, which would mark a sea change for the company and perhaps for the global auto industry as well.

The concept version of the Chevrolet Volt provided information to the driver on a screen-style display that included system details within the speedometer display. *GM*

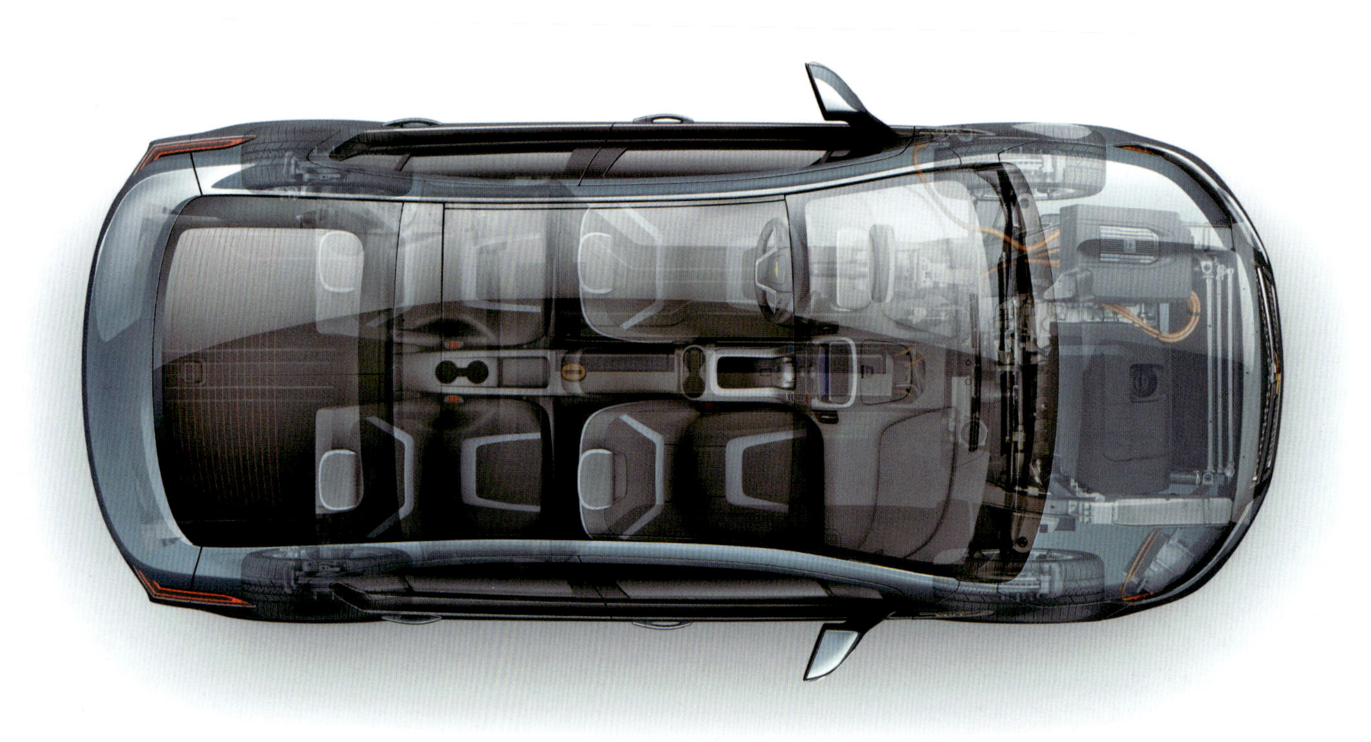

An artist's rendering shows the Chevrolet Volt concept in x-ray view. *GM*

Peel off the bodywork and passenger compartment, and the Chevrolet Volt electric and mechanical components are revealed. The E-flex extended-range electric vehicle propulsion concept was designed to carry either a small internal combustion engine/generator or a hydrogen fuel cell stack to replenish the batteries after 40 miles of driving had depleted the charge they received from being plugged into the power grid. *GM*

In addition to the engineers, the Posawatz-led development team included John Ferris from GM research and development; Courtney Moody, who was with Chevrolet but soon would be joining Lauckner's staff; Cristi Landy from Chevrolet product planning; and public relations staffers from Chevrolet and GM's Tech Center.

With Lauckner, the team had input in crafting a speech GM chairman Rick Wagoner presented a few days after Thanksgiving as the keynote of the 2006 Los Angeles International Auto Show. Though few knew it at the time, the speech would be a precursor for what was to be unveiled a few weeks later in Detroit.

"Going forward, it is highly unlikely that oil alone is going to supply all of the world's rapidly growing automotive energy requirements," Wagoner said in his L.A. keynote. "For the global auto industry, this means that we must—as a business necessity—develop alternative sources of propulsion.

"This is a huge assignment, but it's also an extraordinary opportunity. By developing alternative sources of energy and propulsion, we have the chance to mitigate many of the issues surrounding energy availability. We'll be able to better cope with future increases in global energy demand. We'll minimize the automobile's impact on the environment. We'll be able to take full advantage of the incredible growth opportunity for cars and trucks around the world. We'll take a lot of risk out of our business and likely improve the profitability of the industry. And, not to be overlooked, we'll have the opportunity to make the automobile more exciting, more functional, and more fun to drive than ever before.

"The key, as we see it at GM, is energy diversity. We believe that the best way to power the automobile in the years to come is to do so with many different sources of energy."

Wagoner went on to restate GM's belief in ethanol and other biofuels, but, he added, even flex-fuel vehicles won't "get the job done. Right now, there are about one hundred and seventy thousand gasoline stations in the United States and only about a thousand E85 [ethanol] pumps."

Wagoner announced that "GM is significantly expanding and accelerating our commitment to the development of electrically driven vehicles, *beyond* what we have already committed to with our fuel cell and hybrid programs."

The GM chairman talked about plug-in hybrid vehicles and fuel cell vehicles but reiterated that the company's commitment went *beyond* hybrids and fuel cells.

"GM is committed to the development of electrically driven vehicles that will help improve energy diversity and minimize the automobile's impact on the environment," he said.

Wagoner added that what was ahead was "an unequalled opportunity to really reinvent the automobile" with a transformation "as important as the transition from horses to horsepower."

He promised that General Motors would "follow today's announcements with additional announcements during the auto show season, including Detroit, in about six weeks. There is much more to come from us on this issue. Stay tuned."

GM vice chairman Bob Lutz presents the Chevrolet Volt concept car at the 2007 North American International Auto Show in downtown Detroit and then is swarmed by reporters wanting more details, like if there were any possible plans to take the car from concept into production. *GM*

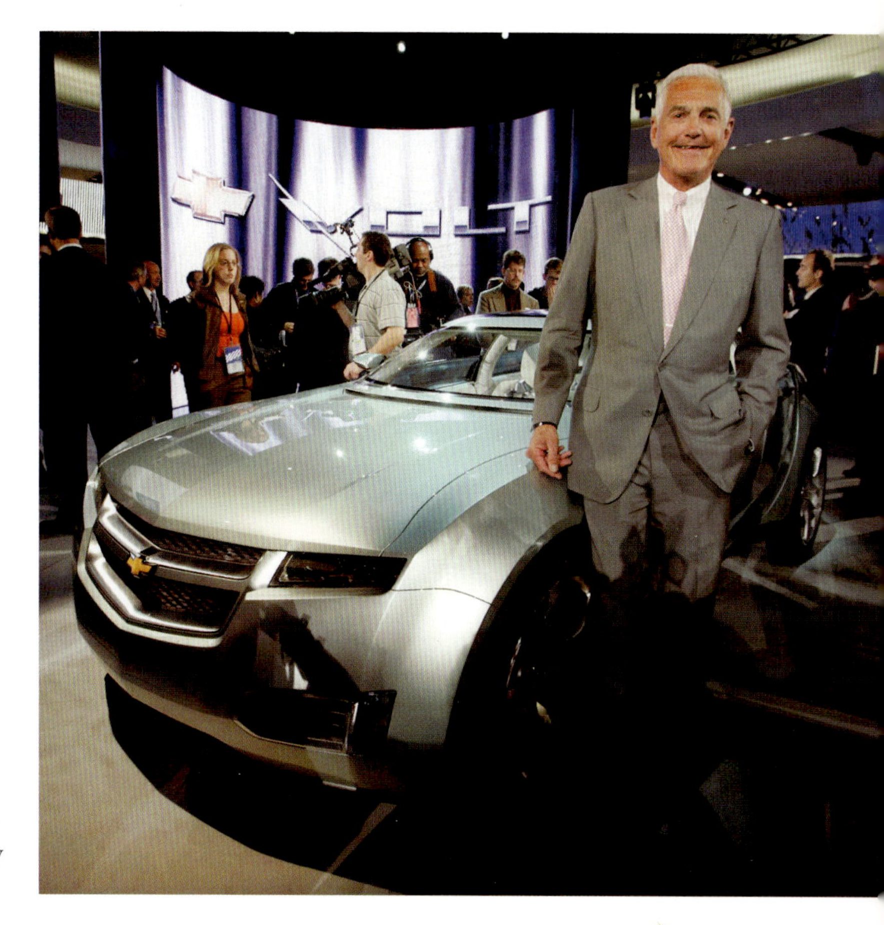

CHAPTER 3

FROM CONCEPT TO REALITY

RIGHT ON SCHEDULE, THE CHEVROLET VOLT CONCEPT CAR WAS FINISHED DECEMBER 22, 2006, AT SPECIAL PROJECTS, A VERY SECURE AND PRIVATE PROTOTYPE CAR-BUILDING FACILITY IN PLYMOUTH, MICHIGAN, A WESTERN DETROIT SUBURB.

BUT EVEN THOUGH THE CAR WAS FINISHED, THERE WAS STILL MUCH WORK FOR THE TEAM TO DO. SOON AFTER GM CHAIRMAN RICK WAGONER'S KEYNOTE ADDRESS AT THE LOS ANGELES AUTO SHOW, THE VOLT TEAM BEGAN A SERIES OF "BACKGROUNDERS," SHARING EMBARGOED INFORMATION ABOUT THE STILL-SECRET CONCEPT CAR AND ITS REVOLUTIONARY NEW TECHNOLOGY WITH AUTOMOTIVE JOURNALISTS AT VARIOUS LOCATIONS IN NORTH AMERICA AND EVEN DOING A SESSION IN LONDON FOR KEY EUROPEAN PRESS CORPS.

The design for the production version of the Chevrolet Volt is officially unveiled as part of General Motors' centennial celebration in September 2008 inside the Wintergarden at the GM headquarters in downtown Detroit. *GM*

"We had had enough discussion among ourselves internally that we knew it would take people time to internalize what we were talking about with this concept," Jon Lauckner explains. "This was not a straightforward concept car—a design in a nice package and a reasonably well understood propulsion system, a gasoline or diesel engine and transmission. This was quite a bit different."

Different indeed. After all, the engine wouldn't drive the wheels. Only the electric batteries would do that.

"We knew a fifteen-minute press conference at the auto show wouldn't be enough time," Lauckner continues. "We wanted to explain for an hour or two what we were doing, the technical description, what characteristics it would have, and to have time for questions and answers so that when we unveiled it, the journalists would be ready to go with their stories."

One story line the E-flex team hoped would be told was how the Volt concept could produce thousands of dollars in fuel savings over time. Yes, new technology is always expensive when it's introduced, but the team noted that the cost per mile driven on electricity would be much less than the cost for traveling that same mile on petroleum, perhaps one-sixth the cost.

Another aspect of the Volt concept was the use of E85 ethanol fuel instead of unleaded regular gasoline to power the on-board generator engine, thus further decreasing the driver's dependence on imported petroleum.

The Chevrolet Volt concept and its E-flex electric drive system were unveiled at the North American International Auto Show in downtown Detroit on January 7, 2007.

"The EV1 'died' because it had limited range, limited room for passengers or luggage, couldn't climb a hill or run the air conditioning without depleting the battery, and [had] no device to get you home when your battery charge ran low," Bob Lutz said. "The Chevrolet Volt is a new type of electric vehicle. It addresses the range problem and has room for four to five passengers and their stuff. You can climb a hill or turn on the air conditioning and not worry about it.

Top: The Chevrolet Volt concept car is featured during the GM style fashion show held in conjunction with the 2008 Detroit auto show. *GM*

Bottom: Actress Katherine Heigl, dressed in Dolce & Gabbana, poses with the Chevrolet Volt concept car at the GM Ten Fashion Show in Los Angeles in February 2007. *GM*

"If you live within thirty miles from work and charge your vehicle every night when you come home or during the day at work, you would get one hundred fifty miles per gallon."

Or, if you used E85 instead of unleaded gasoline, with gasoline providing only 15 percent of the mix, the fuel economy figure became 525 miles per petroleum gallon.

Citing statistics that more than half of all Americans live within twenty miles of where they work, Lutz added that an average commuter "might never burn a drop of gas in the life of the car."

"Today's vehicles were designed around mechanical propulsion systems that use petroleum as their primary source of fuel," added Lauckner. "Tomorrow's vehicles need to be developed around new propulsion architecture with electricity in mind. The Volt is the first . . . "

"Whether your concern is energy security, global climate change, natural disasters, the high price of gas, or the volatile pricing of a barrel of oil and the effect that unpredictability has on Wall Street—all these issues point to a need for energy diversity," added Larry Burns, GM vice president for research and development and strategic planning.

"Today, there are more than eight hundred million cars and trucks in the world. In fifteen years, that will grow to one-point-one billion vehicles," Burns continued. "We can't continue to be ninety-eight percent dependent on oil to meet our transportation needs. Something has to give. We think the Chevy Volt helps bring about the diversity that is needed. If electricity met only ten percent of the world's transportation needs, the impact would be huge."

Journalists covering the Detroit show agreed and so did the show-going public—as well as the General Motors board of directors.

"How General Motors Solves the World's Gas Crisis" read the banner headline across the top of the cover of *AutoWeek* magazine's January 15, 2007, issue.

"If there's a 'Wow, did you see that?' at this year's North American International Auto Show, it is most likely the Chevrolet Volt," wrote Bob Gritzinger, the magazine's senior editor for news. "How, out of some four dozen world debuts, does this unremarkable-looking General Motors concept rate so highly in our estimation? Consider a few facts": he noted, citing the Volt's ability to travel 40 to 45 city miles on electricity, its 640 miles of range from a 12-gallon fuel tank, its ability to use E85 with only 15-percent gasoline content, and its flexibility of employing fuel cells or biodiesel fuels when and where available.

"Not convinced?" Gritzinger continued. "Try thinking of it this way: Most every whiz-bang, high-mileage green car today relies on some form of hybrid powertrain, typically involving a gasoline engine powering the drive wheels, augmented by a secondary electric motor powertrain to provide low-end torque,

Michigan Governor Jennifer Granholm checks out the plug-in cord for the Chevrolet Volt concept at the 2007 Detroit auto show, and then she gets an explanation from Tony Posawatz, vehicle line director for the Chevy Volt (left), and Ed Berry, director of public policy, about the car's electric propulsion technology. The production version of the Volt will be built in Michigan, along with much of its componentry, including the battery packs and even the batteries. *GM*

stop-start function, and some electric-only drive. Volt's E-fleet system turns that on its head . . . "

Praise from the media is one thing. More important was how the General Motors board of directors responded to the concept.

"Right after the auto show, on February 5, we had a discussion with the General Motors board of directors," says Lauckner. "The focus wasn't the Volt. It was more general than that. Basically, it was the whole hybrid strategy, but I did a short presentation about the Volt."

The board's regular meeting was the following day, but on the evening of the February 5 it met in the GM Design Dome where Lauckner led a "walk around" the Volt. He covered the technical concept of extended-range vehicles, the key attributes of the Volt, the impact of 40 miles of petroleum-free driving, how the Volt would lead to almost a plug-and-play architecture for various powerplants. He also mentioned how the team did background presentations before or at the show for some 400 journalists and how the car had received critical acclaim from journalists and the show-going public alike.

"At that meeting, the board encouraged us to get going with a full-scale program to develop a production version of the concept car," Lauckner says.

In anticipation of that meeting with the board, Lauckner made a presentation on January 24 to GM's automotive product board, which included Tom Stephens, Bob Lutz, Rick Wagoner, Ed Welburn, and Jim Queen. Lauckner and the product board anticipated that the board of directors would eventually approve the Volt for production, so they set a timetable for production to begin in November 2010.

That date represented a development process that would be nine months shorter than was typical for a GM vehicle program. And as of yet, of course, there still were no batteries for the vehicle, but everyone agreed that the Volt was too important to the company's vehicle portfolio to wait any longer. To help streamline the schedule, the product board agreed with Lauckner to create a decision-making authority unlike that used for any other GM vehicle. Instead of the usual review process by the GM global product development council, Volt development would be overseen by what was termed "the leadership board": a dozen high-level, U.S.-based GM executives—11 directors and vice presidents, as well as Vice Chairman Lutz.

Lauckner says this new leadership board management group was needed to provide quick feedback and guidance because of the unique nature of the Volt vehicle, because of the accelerated development schedule, and because so much invention would have to be done, especially in regard to the electric extended-range powertrain.

"It was a one-off recognition of the scope and scale of this particular initiative," Lauckner says. "This program was one vehicle, and it had a very specific mission and it required a fair amount of invention. When you have a fair amount of invention, you need a different form of management to keep on an accelerated timeline."

The group also agreed that unlike other GM vehicle development programs, the roles of vehicle line executive and chief engineer would be played by a single person for the Volt. The vehicle line executive's duties involve program management, budget control, and meeting deadlines. The chief engineer's job is to deliver a vehicle that meets all design and engineering goals. But because of the unique nature of the Volt, those sometimes conflicting duties needed the sort of integration provided by someone as unique as the car itself.

Lutz, Lauckner, and Jim Queen, GM's group vice president in charge of global engineering efforts, knew who they wanted to lead the program that would take the Volt from a concept to a production vehicle.

"There are people who like to create things, who create the future, who almost already live in the future," says Frank Weber.

Frank Weber is one of those people. "Things that exist are, for me, boring," he adds with an almost detached-from-the-present matter-of-factness.

Weber, says Lauckner, "was the right kind of guy we needed for the Volt."

"When you have a fair amount of invention, you need a different form of management to keep on an accelerated timeline."

—JON LAUCKNER

Weber, a German, was born into a family that for four generations had been schoolteachers. But there was nothing rote about young Frank Weber's interests or his future.

Weber's passion as a youth was music, though even then he was not interested in what was but what could be. After 15 minutes of formal piano lessons, he was off on his own, teaching himself to play keyboards and writing his own music. He also became fascinated with music produced electronically and built his own radios and organs. He seriously considered studying electrical or acoustic engineering in college.

But he didn't. He found a new passion.

Oh, Weber still plays the piano—and still writes his own music. When General Motors did a series of commercials highlighting the non-automotive aspects of several executives' lives, Weber appeared in one, playing the piano and talking about being inspired.

The inspiration for the expansion of Weber's passions as a teenager was driving. But he wasn't just satisfied with driving cars, he wanted to make them.

"From the moment I was allowed to drive a car and I started working, I started restoring cars," he says. Even more than with the electrical appliances he would fix, Weber found he loved the sound of an engine starting after he brought it back to life.

"I probably spent more time under cars than in the auditoriums [in college]," says Weber, who really didn't ignore his studies in mechanical engineering and now is working on his PhD while he works for General Motors.

Weber's restoration projects included at least 30 vehicles—mostly Volkswagen Beetles and some vintage Saabs. Most were bought, restored, and sold, though he still has the first car he restored: a turquoise Beetle convertible from his birth year, 1966. He also owns a 1953 split-window VW, a barn find that looked on the outside like some ruined relic from an Egyptian archaeological dig. But once Weber removed all the dust and got inside the car, he discovered it had been well-maintained mechanically and that the original owner had covered the seats with vinyl when the car was new; in fact,

General Motors handed the job of taking the Chevrolet Volt from a concept car to a production vehicle to Frank Weber, who brought a unique set of tools to that task. *GM*

he doubted that the cloth had ever been touched by the seat of anyone's pants.

Wanting to make things better than the way he found them became a source of inspiration for Weber's personal lifestyle and for his professional career.

On a personal level, much of that inspiration comes from Weber's wife, an Austrian-born nutritional consultant who showed him how much healthier he feels living as a vegetarian who eats only organically grown food. Their family also disdains the use of air conditioning. Add that to Weber's own environmental consciousness—he notes that energy conservation and recycling are sensitivities you develop growing up in Germany—and, he notes, "You see that things are impacted by your choices. You learn that every small choice you are making as an individual at the end is affecting the system."

In the Volt, Weber found a project where he could make a difference for society, for advancing technology and mobility, and for achieving a goal he believes everyone shares. "At the end," he says, "we all want to leave a better world behind us."

After college, Weber went to work for GM's European division, Opel, in February 1991, in the advanced engineering department where he did concept vehicle development, which, he says, provided him with a technically holistic overview into vehicle creation and systems engineering and their interaction with other onboard systems. "It was a wonderful beginning," he says.

From engineering, Weber moved into product planning, creating scenarios for possible future vehicles and incorporating not only engineering but production and marketing and program management into his resume. Weber's exposure to product planning coincided with General Motors' push into globalization of its engineering programs, thus exposing him to interaction within the company's inner workings on a worldwide basis. He led the planning team that was converging regionally developed midsize vehicles into a single global platform.

"I became the planning director for that midsize platform," says Weber, who was involved in the initial planning and high-level strategy development, the

application of that strategy, consolidation of the architecture, and then the various vehicles spun off that platform. "Not many had followed a whole [global] program through to the start of production," he notes.

Next, Weber returned to his original department at GM Europe, but now as director for European advanced concept engineering, managing a staff of 300 and flying 300,000 miles a year to various points on the GM globe. He laughs when he recalls that the flight attendants on Northwest Airlines knew to order a special fruit plate when "Mr. F. Weber" was on their flight's passenger manifest.

Then, on a Friday evening in February of 2007, Weber took a phone call from his boss at Opel, who said a job was being created that probably would be of interest to Weber.

"Funny enough, I was in Detroit for the motor show in January when they revealed the Volt show car, and I had no idea I would ever be connected to it," Weber recalls.

Weber not only would be connected to the Volt, but he would personify the car, and at the same time the car would be the personification of his own unique blend of experience and interests.

"The reason they approached me was that they were looking for somebody who was tech-savvy and absolutely comfortable in dealing in uncharted territory, who has no problem in creating a new piece of technology and putting it into a vehicle, who has the chief engineer element and also can take the vehicle line executive responsibility, which means all my planning and marketing background, and says I know what customers want and bring the customer perspective," Weber says.

Weber saw this challenge of combining the technical and the commercial responsibilities as the perfect evolutionary step in his professional life—and as an ideal way to affect the evolution of the auto industry itself.

Weber's automotive career got an evolutionary jump-start, he believes, because of those old cars he restored.

"You observe people coming from the university, and maybe they don't have that practical experience," he says. "They approach things so theoretically. They might be very good analysts, but they have no sense, no feel for stuff that is moving, for the workers who have put it together.

"But that experience of doing your own cars, ripping them apart and putting them back together and understanding what it means to provide parts for cars that are thirty years out of date, that sense is still helping me a lot. This practical experience, together with the early engineering advanced holistic thing, is providing me with a skill set that makes it easy for me to talk to all factions of the organization, to talk to everybody.

"I don't work in functional boundaries or silos. I learned that a project comes together successfully only if you are making connections between functions and you are helping in stimulating that process."

"Normally," he continues, "the automotive industry is evolutionary, making smaller incremental steps of improvement. There is a good reason, because customers are not willing, are not forgiving [of big missteps]. They expect the vehicle to work and they expect it to work summer and winter, day and night, and after one hundred thousand miles. Therefore, there is [a] reason why the automotive industry is as evolutionary as it is.

"But sometimes, you see, it requires breaking through that system if there are new opportunities that are developing."

Such opportunities can produce rewards, but they bring risk. Battery development had yet to be done.

And yet, seeing how having the Prius on the road as the first practical hybrid for the typical driver paid ongoing dividends for Toyota, General Motors management was convinced that the Volt had to be the first practical extended-range electric vehicle.

Weber accepted that pressure. "I am not patient," he says. "Waiting for me is painful."

With encouragement from the board of directors, Lauckner went back to the automotive strategy board on February 21. Weber had been selected to lead the Volt development team. By then, decisions had also been made to build the Volt on General Motors' global compact car architecture, using a 1.4-liter, normally aspirated, four-cylinder engine (instead of the concept car's turbocharged three-cylinder) as the onboard, range-extending generator.

The concept car engineering team, led by Tony Posawatz, would expand into the vehicle development team that Weber would lead. Posawatz would become the vehicle line director, second in command to the vehicle line executive when it came to such things as budgets and business and schedule management, and organizing the cross-functional team—much like a sports head coach.

"We had a core team that came from the concept work," Lauckner explains. "Basically, you take the core team and fold that in and staff it in a more significant way, at sufficient levels with the right talent and skill set to make sure we deliver on the big idea." Knowing, of course, that in the case of the Volt, "some of the things we would normally do have to give way to make this successful."

Success was not an option. It was a mandate. The Volt was "one, if not the highest, priority [program] we had," Lauckner says.

Of course, before anything could be termed "successful" in regard to the Volt, there was that pesky matter of the batteries . . .

Opposite and below: The production versions of the Chevrolet Volt draw a big crowd at the Bangkok International Motor Expo in 2008. *GM*

CHAPTER 4

EMPOWERING THE VOLT

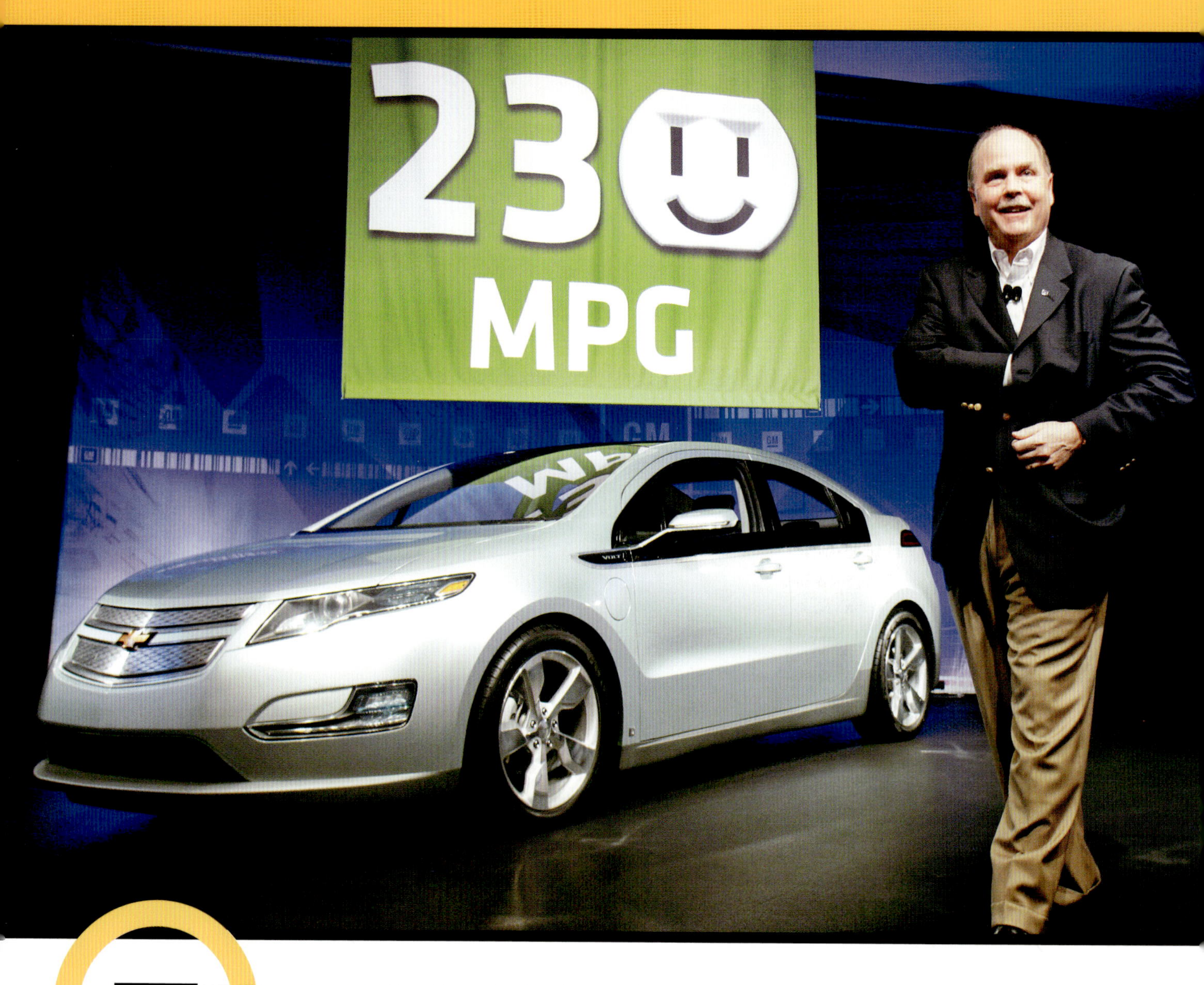

THE IDEA OF PROPELLING A MOTOR CAR WITH BATTERIES IS ALMOST AS OLD AS THE MOTOR CAR ITSELF. THE CONCEPT OF THE SELF-PROPELLED VEHICLE DATES AT LEAST AS FAR BACK AS THE FIFTEENTH CENTURY AND LEONARDO DA VINCI'S DRAWINGS OF AN "AUTO MOBILE" EMPOWERED BY TWO LARGE AND COUNTER-BALANCING DEVICES THAT WE NOW RECOGNIZE AS WATCH-STYLE FLAT SPRINGS. TWO CENTURIES AND AN INDUSTRIAL REVOLUTION LATER, ISAAC NEWTON SUGGESTED THE STEAM ENGINE COULD REPLACE FOUR-LEGGED HORSE POWER USED TO PULL COACHES AND WAGONS.

wasn't until the middle
ıth century that people
rstand it, ancient peoples
the effect of electricity,
Arabic people describing
ıtning strike, the shock
ibed receiving from
e Nile River, or ancient
ıg the force created
rubbed by cloth. (Our
omes from the Greek's
electron.)
cientists in Germany
ds invented electricity-
that Benjamin
ıel as *batteries*, a
rom a "battery of
on description of a
group of large military weapons.

There is evidence that a clay pot battery comprising various metals and citric acid was used for applying precious metals (electroplating) to objects in the Middle East as early as the first century. However, the modern battery wasn't invented until 1800. An Italian biologist, Luigi Galvani, discovered that a leg dissected from a frog would twitch when lightning struck nearby or when the leg received the spark from one of those Dutch or German Leyden jar batteries. Alessandro Volta, an Italian physics professor, replaced the frog's tissues with cloth or cardboard soaked in salt water or sulphuric acid and inserted those materials as separators in a stack of zinc and copper discs. When the top and bottom of this "Voltaic pile" were connected by a wire, a flow of electricity was produced.

Some 80 years later, Volta's name was used to describe the unit of measurement for electricity. Nearly 120 years later, General Motors put his name on its newest concept car.

In between those dates, steam, electricity, and the petroleum-burning internal combustion engine battled to dominate as the primary source for motor vehicle propulsion.'

The authors of *Electric Cars*, a book published in 1977 as part of a science series for Chicago's Museum of Science and Industry, note that in the late 1890s, "the most efficient street vehicles of all were electric streetcars" and that in 1900, "electric-motor technology was far more advanced than gasoline-engine technology."

Many historians have pointed out that in the early years of the motor car, steam, electricity, and internal combustion shared the automotive marketplace. Each had its drawbacks. The fire in a steam car had to be started long before the drive was to begin. Crank-started combustion engines could kick back, breaking fingers or even arms. Electric cars were clean and easy to operate but were slow moving, lacked driving range, and could be used only in areas with electric lines to recharge their batteries and thus couldn't be used in most rural areas.

Early in 2006, a documentary film, *Who Killed the Electric Car?*, was released, criticizing General Motors' for taking back all the EV1s from people who had leased them. But that wasn't

The EV1 was General Motors' electric vehicle showcase at the end of the twentieth century. The Volt shows how far GM has come since then, as it not only has room for more people and cargo inside but uses a small battery pack and supplemental onboard generator to provide range that allows the car to be driven without fear of running out of power. *GM*

Opposite: Fritz Henderson, General Motors chairman, explains how the electric car with extended-range capabilities will achieve the equivalent of 231 miles per gallon by making the most of its batteries and onboard generator. *GM*

The prismatic lithium-ion battery for the Chevrolet Volt (left) was displayed on a table at the battery lab's opening with a cylindrical lithium-ion battery cell developed for the next generation of GM hybrid vehicles. *GM*

the first time GM had, basically, killed off the electric car. Early in the second decade of the twentieth century, Charles F. "Boss" Kettering, who led General Motors scientific and engineering research program for decades, created an electric self-starter motor for internal combustion engines, thus ending the need for and dangers of hand-cranking.

Many electric car companies succumbed. Automotive historian Albert Mroz writes that General Motors ended its own electric car production by 1916, though it continued to experiment with the notion in ensuing years. (There's a photograph in the hallway of the building that houses the GM battery lab that shows a mid-century and rear-engined Chevrolet Corvair that was electric powered by batteries under its hood, and, of course, there was the EV1 that went into production in the late 1990s.)

For most of the history of the automobile, the source of electrical power for starting the engine and supplying the various electric accessories has been the rechargeable lead-acid battery, originally created in the 1850s by French physicist Gaston Plante. Composed primarily of lead, lead dioxide, and sulfuric acid, the lead-acid battery has become familiar to vehicle owners as the heavy, rectangular box topped by two terminals that is located adjacent to the engine under a vehicle's hood. General Motors' EV1 electric vehicle was powered by 26 such batteries.

German scientist Carl Gassner created the "dry cell" battery in 1887. Gassner's zinc-carbon cell was further developed over the years, leading to the work by Lewis Urry for Eveready (Energizer) and the introduction of the alkaline battery in the mid-1950s. Two decades later, the nickel hydrogen battery was developed for communication satellites, and several years later consumer-grade nickel-metal hydride batteries were introduced. These can be found in gasoline-electric vehicles, such as the Toyota Prius, Ford Escape, and hybrid versions of several large General Motors trucks and sport utility vehicles.

While such NiMH batteries can supplement the power produced by an internal combustion engine, they have not proven practical for powering an entirely electric-propelled vehicle, such as the Volt, which has no direct

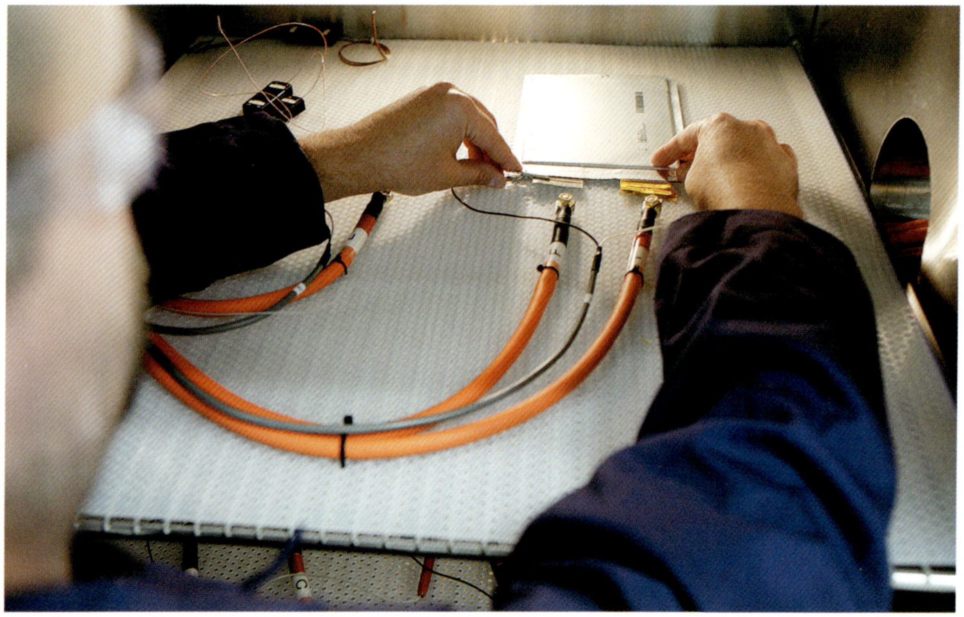

Jack Laerzio sets up a test of lithium-ion batteries in one of the test cells within the GM lab. *GM*

mechanical link between its onboard engine/generator and the driven wheels.

Because of its low density, lithium has wonderful electrochemical potential. Scientists began experimenting with lithium batteries as early as 1912, and lithium-based batteries were on the market in the 1970s. In the 1980s, John B. Goodenough, an American chemist working at Sony, led a team that produced a rechargeable lithium-ion battery, which is in widespread usage in many portable electronic devices, such as laptop computers and cellular telephones.

As early as 2003, California-based electric vehicle development company AC Propulsion built the tzero, a two-seat electric roadster prototype powered by 6,800 lithium-ion laptop batteries. Silicon Valley–based Tesla used that 6,800-cell architecture for its similar-sized roadster.

In 2006, in the Chevy Volt's missing piece of the puzzle was a lithium-ion battery that could provide the performance needed for a mainstream sedan, which could comfortably carry four (or perhaps even five) people and their luggage through all sorts of traffic environments and weather extremes.

"I did a *1-800-I-need-help* call," laughs Denise Gray, who would be responsible for finding those batteries and making sure they could be packaged into packs that would fit into the Volt chassis.

In the summer of that year, while work was underway on the Volt concept, Gray had been promoted to become General Motors' first global director of rechargeable energy storage systems. Her job included overseeing the development of batteries for the Volt and all GM hybrids, as well as future battery technology strategy.

Gray had grown up in Detroit. A middle school teacher noted her excellence in science and math and suggested she consider engineering as a career. Responding to such encouragement, she enrolled at Cass Technical, a magnet high school where she was one of two girls in her class of 35 students.

"It became clear this wasn't a common field for girls to go into," she recalls.

Gray soon found herself in the school's co-op program, spending half a day in class and half a day working in industry. For her, that industry turned out to be the General Motors Technical Center. After high school, she enrolled at GMI, the General Motors Institute, now known as Kettering University. She graduated with a degree in electrical engineering just as cars were becoming more and more computerized.

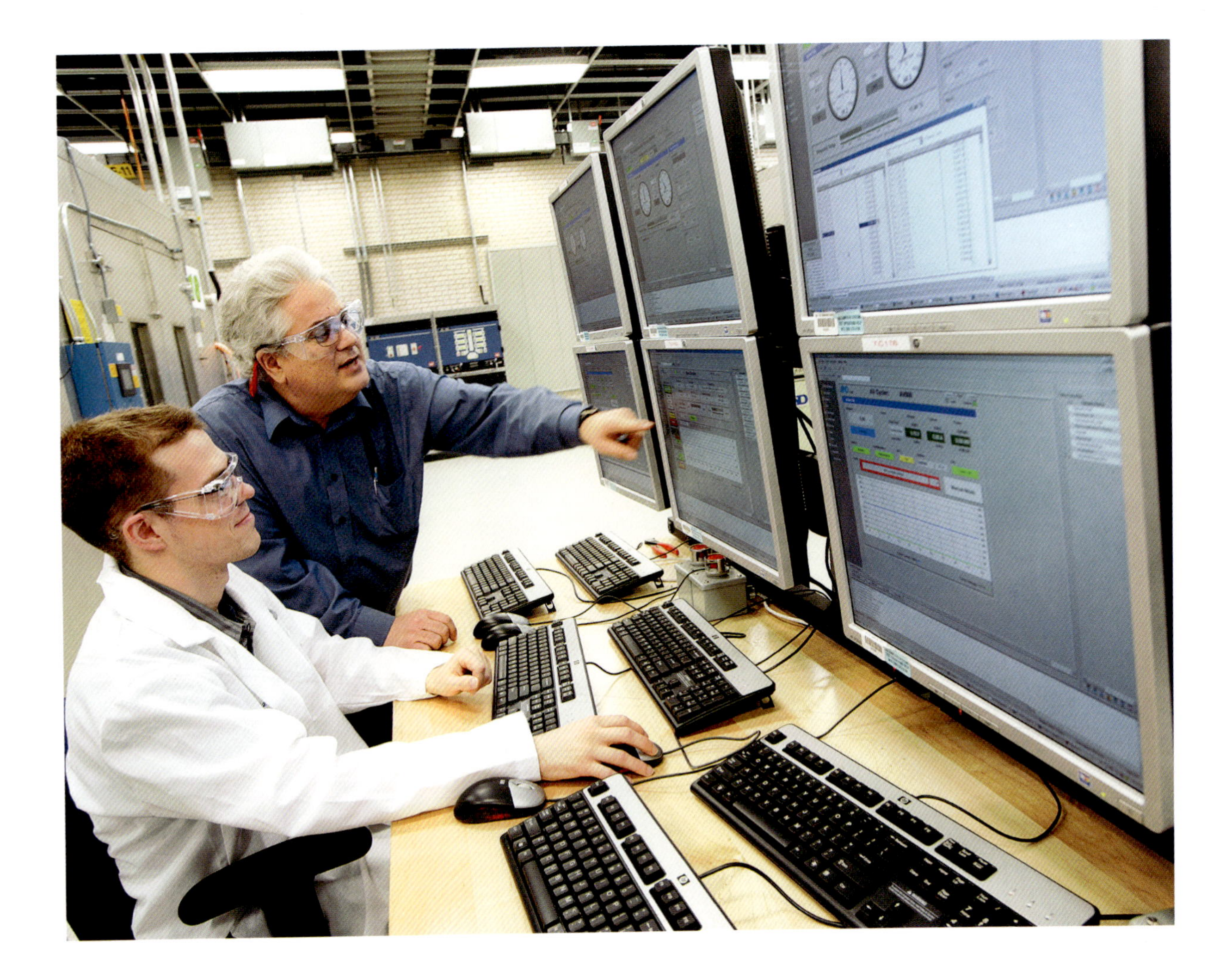

General Motors' battery lab is filled with various testing chambers, all of which can be monitored from various computer screens. Testing runs around the clock, with the computers programmed to call engineers at home if there's an issue that needs immediate attention. *GM*

Lithium-ion battery cells are groups together in T-shaped battery packs that provide power for the Chevrolet Volt. Like the cells themselves, the battery packs undergo extensive testing in the GM battery lab. *GM*

Not only are battery packs tested in the lab, but so are systems that interact with them, including those designed to keep the batteries cool and operating at optimum capacity despite a wide range of ambient temperatures. *GM*

Denise Gray led the battery development and testing program for the Chevrolet Volt. *GM*

As a full-time GM engineer after graduation, she worked in vehicle electrical systems and then in electric controls and propulsion system software that managed increasingly complex electronic systems. In the 1990s, she helped build a team that brought the development of such systems in-house instead of being dependent on outside suppliers.

"Bringing it in-house really allowed us to optimize our propulsion system control systems—[for] engines and transmissions—to make those systems very robust and very cost effective," she says. "I was part of that."

And now, in her new role, she would be a crucial part of the Volt's viability as a production vehicle.

When Gray's new position was created, General Motors already had launched a hybrid version of the Saturn Vue and was readying to expand such offerings to include the Chevrolet Malibu, full-size pickup trucks, and large sport utility vehicles.

As with other manufacturers' hybrids, "that was all [in the] nickel-metal hydride arena," Gray says. But, she added, GM also had been studying lithium-ion options for a couple of years in anticipation of such batteries becoming ready for automotive application.

For the Volt concept, "we had done some work on paper to determine we [could] package it, as well as some work on the technology readiness, but we hadn't actually built one [a prototype running off a lithium-ion battery pack]," she adds.

After the concept was unveiled and was propelled into the pipeline for potential production, "in the ensuing months that was the task: to create one, to finish up the initial design, to quickly get hardware in so we [could] determine [if it could] be realized and where were the areas where we really had to concentrate. By June we had actually gone through a combing of the globe, if you will, with battery companies and technology companies that would want to partner with us to develop that."

Some two dozen lithium-ion battery makers sought the Volt contract.

"We evaluated about twenty-five different cell chemistries and constructions," says Bob Kruse, who, as executive director of global vehicle engineering for hybrids, electric vehicles, and batteries, was in charge of labs where General Motors would test potential Volt batteries and battery packs, validating, among other things, that such devices would warrant the decade-long warranties expected for automotive electric propulsion systems. His domain also included Building 16 at GM's Milford (Michigan) Proving Grounds, where dynamic testing activities for hybrids and battery-powered vehicles was based.

Of those two dozen cells, "we picked two: the A123 cell and the LG cell," Kruse adds. "Then we evaluated them for all of 2008."

"The cell technologies were purposely picked to be different because we wanted to keep our options open," Gray says, adding, however, that the cells from A123 and LG Chem "were the two promising technologies that we wanted to continue to explore."

"The cell technology is really where the stretch is," she adds. "The rest of the system is conventional components, although adapted to the [specific] battery. The main activity was in the cell technology."

In June, General Motors announced contracts with Compact Power Inc., a Michigan-based subsidiary of Korean battery maker LG Chem, and with Continental Automotive Systems, a division of German auto supplier Continental AG. Two months later, cells from Massachusetts-based A123Systems became part of the proposed Continental package—A123 would do the cells and Continental the battery packs.

"A123Systems and LG Chem are both top-tier battery suppliers with proven technologies," Gray said at the time. "We're confident one, or possibly both of these companies' solutions will meet our battery requirements for the E-flex system."

By October, GM had, as Gray puts it, "hardware in-house" and began testing cells and packs in the laboratory.

The testing of the original two-dozen candidate cells was done in the battery development lab GM had created for the EV1 electric car program. Evaluation of the hardware from the two finalists was done in that lab as well, though in response to Denise Gray's "1-800 call," GM's fuel cell engineering group became involved. It already was working on hydrogen fuel cells as a second source for the electricity to power a future version of the Volt but became more immediately involved by offering its technical expertise to help produce early battery packs so Volt vehicle engineers could begin dynamic development work with prototype vehicles.

"Lithium-ion is the reason we can do a Volt," Denise Gray says.

"Lithium-ion is the reason we can do a Volt,"
—DENISE GRAY

Prabhakar Patil agrees. Patil is chief executive officer of Compact Power Inc., which emerged as the battery supplier for the Chevrolet Volt. Before becoming one of the first employees of LG Chem's American subsidiary in 2005, Patil, who has a PhD in aerospace engineering from the University of Michigan, had spent 27 years at Ford Motor Company, where he was chief engineer for hybrid technologies and for Ford's first hybrid vehicle, the Ford Escape that was introduced as a 2004 model.

"I would have liked to have put in lithium-ion [in the Escape] because it has advantages in terms of weight—you can cut the weight by a factor of two compared to nickel-metal hydride—and the size by fifty percent," he says of the Escape hybrid. "But there were serious safety issues, and there was obviously no way you were going to put something like that in a car."

"So," he laments, "I threw out lithium-ion and went with nickel-metal hydride," as did other hybrid vehicle producers.

And what were the safety issues with lithium-ion? Do you remember all of those laptop batteries that were recalled after several portable computers burst into flames?

"Typically, inside the cell there is a positive and a negative electrode, and you connect them with a wire and that's what makes the juice flow," Patil explains.

"But the electrodes have to be kept separate because if they touch each other, it's all over because the current would rather go inside rather than go to the outside, and [if that happens] it rushes so fast it becomes very, very hot.

"So what can happen, and what did happen in some of the laptop recalls, is there was some manufacturing debris left inside the cell, real tiny particles from when you're welding [on] the cap, and it didn't do anything bad initially, and therefore it escaped end-of-[assembly]-line inspection. But as the cell aged and cycled, it went through this compression and release cycle and the debris works its way around and eventually can penetrate the thin membrane that goes between the two electrodes. It's a pinpoint prick, but it's enough that it creates a very intense hot spot.

"Once that happens, the membrane shrinks and it melts and the hot spot gets bigger and the electrolyte that is used in this cell—it's an organic solvent—evaporates . . .

"On the cathode side, people use different materials. . . . Some of those materials can give out oxygen when they get hot, so you have a source of heat and something that can evaporate and burn. Now you have something that supplies oxygen, and you have the perfect combination and the cell takes off." On fire, of course.

"However, by the time I came here, in 2005, things had changed and one of the things that we—LG Chem/CPI—had overcome was some of the safety issues. That's not true generically for all lithium-ion; you have to have unique chemistry of the things you do inside the cell to ensure it's going to be safe."

There also were things LG Chem had to do to make its safe lithium-ion batteries suitable for the needs of a vehicle like the Volt. Until the Volt, the target vehicle for lithium-ion batteries had been hybrid vehicles in which an internal combustion engine provides most of the propulsive force. The pure electric range for such vehicles, Patil notes, is less than a mile. "It will start in a pure electric mode and it can go short distances, but not for long distances, so the energy content, which is what determines the range, was, by design, relatively low, and that's how the chemistry was developed—to give it a lot of power but not that much energy. Those are the cells that were very close to production."

But General Motors wanted the Volt to be able to travel 40 miles without the onboard generator engine running.

"Now the Volt required a different kind of a cell because forty miles of pure electric range meant it had to have a lot of energy, and we did not have [such] a cell on the shelf. That's where our saga began."

"Basically everybody who's anybody in the battery business bid [on the Volt project] because of the high-profile nature of the vehicle and the reputation and recognition it would bring," Patil says.

The challenges included packing enough power into the cells, fitting those cells into packs within the space defined by the Volt's design, and making sure those cells would operate in a wide range of environments, and do so for at least 10 years. Patil says it was a daunting challenge made even more complex by the fact that General Motors had set a November 2010 date for the start of Volt production.

"That was a very, very compressed time frame even if you had a cell on the shelf," he says. "And here we were, starting with no cell on the shelf, and the pack was going to be completely new because the size of the battery that was needed to provide forty miles of range and the space was dictated because of what was available in that T [shape], and the vehicle size wasn't going to change. The space available for the battery was fixed. The range was fixed. So you had to fit the battery in and provide the energy.

"That's when we had to go to some unique things."

Those innovations included liquid cooling of the batteries.

"The temperature is the most critical factor for determining battery life and that means the battery has to be cooled," Patil explains. "A cell phone or laptop battery typically goes where people go, so the temperature environments aren't extreme and, secondly, the life is two years because you're throwing away the cell phone at the end of that time. In a car, that life is ten years. To guarantee that kind of life and to be able to live in the temperature extremes of minus forty degrees to plus forty degrees—Celsius—and going from the summer in Arizona or Las Vegas to the winter in Detroit and the northern states, you had to have a thermal management system."

For typical lithium-ion battery use, Patil says, the battery can be air cooled simply by making the cell 40 percent larger to allow for airflow to provide the cooling. But making the cells large enough for air cooling wouldn't work for the Volt because of the predefined space available for the battery pack.

"We had to have a more efficient way of cooling," Patil notes.

"Of course," he adds, "being electrical entities, the cells don't like liquid and especially water, something that conducts electricity, so they had to be jacketed; the liquid had to flow through some sort of jacket, or fin, as we call it. That was another challenge. That whole liquid-cooled pack technology had to be developed.

"Those were the dual challenges for developing a new cell and getting it proven for high-volume production. The Volt uses between two hundred and four hundred cells, and when you have fifty thousand units of volume, you're talking about a lot of cells. The cell count gets into the fifteen to twenty million kind of range. And you have to produce them consistently because your pack is only as good as your weakest cell.

"You not only have to prove the cell capability in a sample size of one or two or a handful, but you have to have the high-volume manufacturing capability to be able to make millions of them very consistent with each other and at an affordable cost. That was the challenge we had to take on the battery side—developing the cell, making sure it could be done consistently in high-volume production and in pack technology that had to involve liquid cooling."

The LG Chem/CPI solution was to use manganese-based cell chemistry for the cathode, rather than the usual cobalt-based or even the phosphate-based chemistry used by some competitors (Patil explains that manganese-based cell chemistry greatly reduces the release of oxygen should a cell ever overheat) and to develop a unique ceramic-coated separator within the cell (thus reducing the potential for penetration by a factor of five or six and eliminating the possibility of melting or shrinking).

"So you have two levels of protection built inside the cell," Patil says, adding that "we also put it through all kinds of abuse testing that a normal customer would never do, including crash and crush and deliberate overcharge, and convinced ourselves that it is absolutely safe."

Patil says the work began in April—with cell chemistry handled by LG experts in Korea and battery pack development done in Michigan at Compact Power. The deadline for delivery of the first battery pack to General Motors was—trick or treat—Halloween.

"It took a lot of working through the nights and on holidays and whatever, but the first pack, the prototype pack, was delivered, and it's been on the test bed and still is on the test bed and has been performing," Patil says, adding that since that first pack, LG Chem and GM have worked together to improve the batteries and pack performance in the months leading up to the start of Volt production.

The battery pack and the car designed to carry it. *GM*

"It was literally taking the technology leadership back to this country or back to General Motors, and that was something that was very important, as well as what it does for the environment."

—PRABHAKAR PATIL

He notes that both companies knew the process would be difficult but had more in mind than just completing the project. "[We told General Motors] we believe in the cause and what it will do, and, by the way, that cause is something that drove our team as well because it wasn't just business," he says. "It was literally taking the technology leadership back to this country or back to General Motors, and that was something that was very important, as well as what it does for the environment."

"But we recognized the risk. There could be things that happened that none of us had any control over that would set us back. As long as we jointly recognized that—on the vehicle side and on the battery side—we would work together and do everything we could."

Patil says he was overwhelmed by General Motors' willingness to use a teamwork approach. "There were as many GM engineers here as there were our engineers," he says. "They just happened to have different badges. But we worked together, full transparency."

Patil says one thing in favor of the LG/Compact Power effort was the extensive automotive background shared by Compact Power's employees. (Many came from automotive backgrounds.)

"We know the technicalities of the gateways [automotive development waypoint deadlines], the timing you have to meet, the quality that is required, the consistency that has to be there, the things people like General Motors and Ford appreciate. For me [at Ford], one of the things that was frustrating was dealing with a non-automotive kind of supplier because people don't understand that if they come from consumer electronics or even aerospace, the kind of rigor it takes to work in the automotive business. Having people on the battery side who understood that helped a lot and avoided a lot of wasted effort and the finger-pointing that can go on."

Major milestones in the process, Patil says, included proving the thermal management system worked, helping GM develop an accelerated testing procedure so batteries could be exposed to 10 years' worth of cycling in a 2-year period, and devising battery packs so their assembly process was amenable to mass production without becoming too expensive and at the same time would be strong enough to withstand crash-testing procedures.

"We were really thrilled when we were selected out of that contest because that was basically being recognized as getting GM's seal of approval."

Ann Marie Sastry, the Arthur F. Thurnau Professor of Engineering at the University of Michigan, speaks at the opening of the GM battery lab. GM helps fund battery research and engineering education at the university. *GM*

On June 13, 2009, the first completed and tested battery pack is inserted into a Chevrolet Volt IVER (integration vehicle engineering release), a production-intent prototype that will be used for dynamic testing and validation in preparation for the production of the 2011 Chevrolet Volt production car. *GM*

General Motors announced at the North American International Auto Show that it had awarded the Volt battery contract to LG Chem/Compact Power Inc. GM also announced at that show that it would become the first major automaker to build its own battery packs and would join with the engineering school at the University of Michigan to do advanced battery research and establish a specialized graduate-level curriculum to develop automotive battery engineers.

In the fall of 2009, Compact Power announced a $300-million project to begin building lithium-ion batteries in a new plant in Holland, Michigan, in 2012. Until that plant is completed, batteries for the Volt come from Korea.

Patil notes that until General Motors' own battery pack plant became operational early in 2010, Volt battery packs were produced at Compact Power's Michigan facility.

"General Motors made a decision, from a strategic point, that they wanted to build the packs," Patil says. "Even though we wanted to do the packs, and were getting suited up to do the packs for volume production, we understood and split it apart and agreed that General Motors would go ahead. It's a joint design and development, and we have the capability to make the packs here if required."

"When we started down this path, we envisioned that we would be buying packs," says GM's Bob Kruse. However, "along the way, we recognized that we were going to put a heck of a lot of resources and effort into making sure the pack was successful, so we decided—it was a strategic move—to design and manufacture the pack in-house at General Motors and to buy the cells from LG Chem."

Kruse notes that of the 150 unique parts that compose the T-shaped battery pack that powers the Volt, all but 8 are designed and engineered by General Motors.

"The battery pack is a lot more than just a collection of cells," he adds. "There's a lot of electronics and thermal control systems.

"How you treat those cells, to get them to last the life of the vehicle, you don't over-charge them, you don't over-discharge them, you don't do any of that at hot or cold temperatures. You treat them right to get them to last a long time. That's intellectual property that's very unique to General Motors."

"Quite frankly, we felt we could go the fastest if we could bring it in-house," adds Denise Gray, noting General Motors' more than a century of experience in manufacturing, materials, and structural design. She adds that the

A cutaway of the chassis shows the position of the various Chevrolet Volt powertrain components, including the placement of the lithium-ion cells within the battery pack that runs down the center of the vehicle and beneath the back seats. *GM*

A small four-cylinder engine generates electricity to recharge the batteries, but doesn't turn the car's wheels as in a traditional car. *GM*

automaker also had extensive expertise in creating complex electronic and thermal control systems.

"For example," she says, "you have to understand electrochemically what's happening [in each cell], what's the signature, what's the temperature signature, is it uniform across [the pack]?" You also not only have to monitor such things but devise analytical models and software tools and testing hardware, so, for example, 10 years' of use could be simulated in the lab in the 2 years her team had to prepare for the start of production.

Bringing that work in-house, she says, likely cut a year off the process and thus preserved the Volt's very compressed development time schedule.

Early in 2009, General Motors opened a $28-million, 33,000-square-foot, state-of-the-art Global Battery Systems Lab in what formerly was the GM Performance Division (motorsports) area of the Vehicle Performance Center, a building on its Technical Center campus in Warren. Alternative Energy Center was the new sign on the front of that building, which is located just behind the company's huge and primary Vehicle Engineering Center.

"We tested every single battery pack in our lab before putting them into our [development] vehicles," Gray says. "The battery development was done in parallel with the vehicle development, and we couldn't let vehicle development stop because the battery stopped. We tested all packs to ensure they could do their work."

General Motors Global Battery Systems Lab includes 32 "cyclers" (think of them as treadmills) for battery pack testing and 32 more for cell and module testing. Pack cyclers can test 64 channels of information; cell cyclers test as many as 96 channels. There also are 25 thermal chambers for pack testing and 16 for cell testing. In addition to being tested at a wide range of temperatures, the cells can be adjusted to varying levels of humidity, and there's a test chamber that puts packs into a chassis and shakes and vibrates them to extremes.

"We can make it hot or cold. We can cycle it and shake it. We can cycle it and shake it and bake it all at the same time," says Bob Kruse.

The effect, he says, is to simulate in the lab a decade of driving in one fifth of that time. The lab also is used as development continues to future generations of cells and packs for future Volts and other E-flex vehicles.

Though engineers—some of whom brought experience from working in the lab during EV1 development—are on hand for only one work shift, testing goes on around the clock, with computerized monitoring that phones an engineer at home if it detects any fault. The engineer can log in from home to see what's happening. "They can stop a test or shut down the whole lab and come in," says Kruse, after many months of testing had been done, adding, however, "but we haven't had to."

The lab was built not only with battery testing in mind but to reflect the Volt's environmental benefits. Low-voltage, low-wattage light-emitting diode lighting was used throughout. Floors are made from recycled tires. There are electricity-generating wind turbines on the roof, and the lab sends power back to the electrical grid.

"We know we're making history," says Kruse. "Fifty years from now, everybody will remember the Chevrolet Volt like they remember a '53 Corvette. We tend to remember iconic vehicles. It's very clear the Volt will be iconic. The men and women who are working on it know they are making history."

"Electrification of the automobile is a big deal," he added. "We're fortunate to be at this time in history working on it." Fortunate, but also facing significant challenges, it seems.

Prabhakar Patil of Compact Power remembers how he used to remind his team when it felt under pressure that it was charged only with dealing about the battery."Think about General Motors," he said at the time. "[It is] dealing with the whole vehicle."

Patil was the chief engineer for Ford's first hybrid vehicle and understood the challenges of dealing with the whole vehicle. Yes, he had a pretty good idea of the magnitude of the challenges facing the Volt engineering development team.

Jim Queen, group vice president for global engineering, speaks at the opening of General Motors' 33,000-square-foot Global Battery Systems Lab at the GM Tech Center in Warren, Michigan. The laboratory became a hub for the reinvention of the automotive powertrain, with the Chevrolet Volt leading the way toward practical electric propulsion. *GM*

CHAPTER 5

INSULATING THE VOLT AGAINST THE "ANTIBODIES"

For Gery Kissel, the signal that General Motors truly was serious about the Chevrolet Volt didn't involve its unveiling at the 2007 North American International Auto Show, nor was it clear from the early testing of potential lithium-ion battery suppliers. For Kissel, the signal that the company was serious came in August of 2007 when it brought Andrew Farah back from Europe to be the chief engineer for the effort to take the Volt from a concept car to a production vehicle.

Kissel had been a General Motors engineer for more than a decade. With degrees in electrical engineering from the University of Michigan and Purdue University, he had become a high-voltage wiring and safety specialist for GM and had worked on electrical interface verification on the EV1 electric car, on various hybrids, and on GM's futuristic fuel cell concepts.

Kissel was working on a plug-in hybrid vehicle project when the Volt concept was unveiled. He saw the very positive early reaction to the car and saw momentum building within parts of the company to take the concept to production. But what convinced him that the Volt would, indeed, become a viable vehicle was the hiring of Farah to oversee the day-to-day engineering effort.

"This is the right person to really make this happen," Kissel remembers thinking at the time. "Andrew worked on the EV1. He had the vision for electric vehicles. When Andrew was picked, for me this was really going to happen."

And yet, at first, Farah declined the offer to end his assignment in Europe early and to come back to the United States as the Volt's chief engineer.

Here's how he remembers getting the call from Detroit: "We have a new job for you," they said.

"What is it?" he responded.

"We're going to make an electric car called the Volt."

"Not interested."

"What do you mean?"

"I don't think we're serious about it," he said. "We talked about this idea for years. Why suddenly now?"

Farah was assured the effort was genuine and backed by the highest levels of GM management. He still wasn't convinced. He insisted on talking to those within the company who could relieve his doubts. He flew back to the United States for those meetings. Not only was he convinced of the company's commitment, but he immediately realized the powertrain configuration would fix many of the ills that plagued the EV1.

In fact, he says, he remembered a similar powertrain proposal being made back in 1994 when the EV1 was being developed and again later, just after the program was cancelled.

Farah was one of the development engineers on that program and was involved in creating a rudimentary form of a range extender for that electric vehicle.

"We were driving the EV1 back and forth across the state on a regular basis [during development]," he says. Because the engineers needed more miles to do their work than the batteries alone would provide, the engineers devised small trailers equipped with gasoline-powered generators that their EV1 test vehicles towed along behind. "Push a button and it generated electricity, and as long as you weren't driving faster than sixty miles per hour, you could keep driving until the gas ran out. It was a very similar concept to the Volt."

Farah remembers being in meetings with Frank Weber before Weber was called to the United States to lead the Volt program and remembers being a little jealous of Weber's new position. "What a great job that would be," he remembers thinking, of course, if GM truly was serious about the program.

Turns out, the company was, so Farah called his wife and told her that as much as they had enjoyed living in Europe, they were returning to the United States.

Farah began working for General Motors as a college student, while he was studying computer engineering at the University of Michigan.

"At the time, the idea of computers in cars was just taking off," Farah says. "They didn't know what to do with a computer engineer, so I initially worked in the computer department writing small programs that sorted data. I was working the third shift, putting paper in printers and printing payroll checks."

After graduation, Farah helped do testing and validation of networked microprocessors in the GM30 cars and worked on the TV-style information display screen that debuted in the 1986 Buick Riviera.

Chief engineer Andrew Farah wasn't worried about the General Motors bureaucracy as far as the Chevrolet Volt (opposite page) was concerned, but he knew such a radical vehicle could be vulnerable to what he called the "antibodies" within the GM engineering culture. *GM*

Below: Chief engineer Andrew Farah had suggested a Volt-like vehicle when he was part of the team that developed the EV1. *GM*

The team at prototype-building specialist Special Projects creates the Chevrolet Volt concept car from plans created by the GM Design staff. *GM/Will Handzel*

Even with computer-aided engineering math models, creating a concept car is an incredibly labor-intensive and time-consuming task. *GM/Will Handzel*

Every part of the concept car is created by hand and from scratch. *GM/Will Handzel*

He continued to work in product development on projects that brought new technologies into the industry but later left GM to work for Johnson Controls, a major automotive supplier of batteries that was expanding into complete battery systems. Economic circumstances led to cuts in his department, and just as it looked like he might be out of a job, he ran into a former GM coworker (Jon Bereisa) at an engineering conference and learned that GM was thinking about doing an electric vehicle.

"I became the second person in the propulsion controls and integration area and eventually managed that group," Farah says of what would become the EV1 development program. At least twice, Farah says, he proposed incorporating the engineer's range-extending generator trailer into the vehicle, but "I was shouted down by the purists who said, 'No, this is a zero-emissions vehicle.'" After the EV1, and with no similar vehicle on the GM horizon, Farah became vehicle line manager, a more business-oriented position, on the GM W car (the Chevrolet Impala and Monte Carlo and Buick Century). Other assignments followed until he went to GM Europe to lead after-sales engineering, dealing with service, parts, and accessories across Europe.

Nick Zielinski, the chief engineer for development of the Volt concept car, had a list of things he needed to do before he could hand the project off to the production chief engineer. However, the Volt program was, as Farah phrases it, "a rocket sled." Meaning? "I basically just moved in right across the aisle from [Zielinski], and we both went at it until we reached the VPI [Vehicle Program Initiative] gate [one of several milestones in any General Motors vehicle development project], which must have been in April of '08."

The Volt, Farah notes, "was violating all the [normal GM vehicle development] rules." For example: "All the technologies should have been proven to a point that you're ready, and we were violating that in spades. It created a lot of process problems for the broader organization."

Fortunately, Farah says that early in the program, decisions had been made that would shelter the Volt from what he

A previous-generation Chevrolet Malibu body shell is prepared for its conversion into a prototype "mule" that will be used to validate various powertrain components designed for the Chevrolet Volt. *GM/Will Handzel*

Various components are laid out in preparation for their installation into a Volt mule. *GM/Will Handzel*

termed the organizational "antibodies." Antibodies are not the same as corporate bureaucracy, he explains. He compares dealing with bureaucracy to driving through heavy snow. It can make the journey difficult and bury you for a while if you stop your forward progress, but if you persevere, you eventually reach your destination.

On the other hand, antibodies actually attack you. Stop and take a breath and they swarm you.

General Motors, Farah says, is very skilled at making "many, many, many incremental changes." But we aren't used to, aren't comfortable with, making "a significant number of significant changes," such as had to be done with the Volt and its revolutionary powertrain. That's when the normal processes break down and the antibodies kick in.

"If this program was treated like a regular program, it would fail. Simply put," Farah says. "Because all of what I refer to as the antibodies of the organization would kick in and kill it, either early in the process because of certain business cases or performance situations or whatever; the antibodies would just come in and kill or delay the thing."

However, he notes, decisions were made early on, decisions like establishing a unique management board and putting people in management roles

Top: GM engineers and technicians prepare a battery pack for its installation into a "MaliVolt." *GM/Will Handzel*

Bottom: The battery pack is lifted into its place beneath the MaliVolt chassis. *GM/Will Handzel*

because of their technical expertise or creative vision, not because they'd been, for example, chief engineer on many other vehicle programs.

"There are a lot of folks working [on the Volt] where there are no templates, where there is nothing that has come before them. Or if it did, it might be the EV1, and the EV1 was a different vehicle. There are some similarities, but still, it was a very different thing and not everything applies," Farah says.

The antibodies' reaction would be, "we can't do that," Farah says, while his approach was saying, "I have to do it," or maybe, "Who's going to get it done?

"To me, everything is possible," says Farah.

About the time Farah arrived, three previous-generation Chevrolet Malibu sedans had been outfitted with prototype Volt powertrains—a lithium-ion battery pack and a small, internal combustion generator engine so preliminary testing of the drive system could begin at General Motors' Milford Proving Grounds.

It was about this same time that Jon Lauckner had another meeting with the General Motors board of directors.

In February 2007, the board agreed to preliminary funding for engineering development. That August, Lauckner provided a progress report: Frank Weber was on board. Batteries from two potential suppliers were being evaluated. The global compact car architecture would be used—and the board was shown photographs of early styling proposals. The on-board engine/generator would be an American four-cylinder rather than the European three-cylinder in the concept car. A Volt with a hydrogen fuel cell instead of the internal combustion generator also was being laid out for future possibilities.

"We outlined where we were in terms of the key requirements and talked about the challenges that had to be overcome," says Lauckner.

Because it was important not only to produce such a car but to be the first to market, there was agreement that if the Volt went to production, cars would begin making their way along the assembly line in November 2010, pulling that date forward by some nine months compared to the typical GM vehicle development program.

Indeed, Farah and his team had challenges to overcome.

As Stuart Norris (leader of the "human machine interface" design group) discovered, the Volt program involved a unique group of people. "They are passionate about electric vehicles, and, my God, they know about electric vehicles beyond knowing," he says.

"A lot of people on the Volt were people who worked on the EV1. They have electric vehicle just running through their blood."

In June 2008, the General Motors board of directors voted to inject a transfusion of hundreds of millions of dollars into that bloodstream. The board approved the Chevrolet Volt to go forward into production.

A gasoline-fueled-generator engine is maneuvered into place for installation in a MaliVolt prototype vehicle. *GM/Will Handzel*

Myriad connections are made after the generator engine is in place. *GM/Will Handzel*

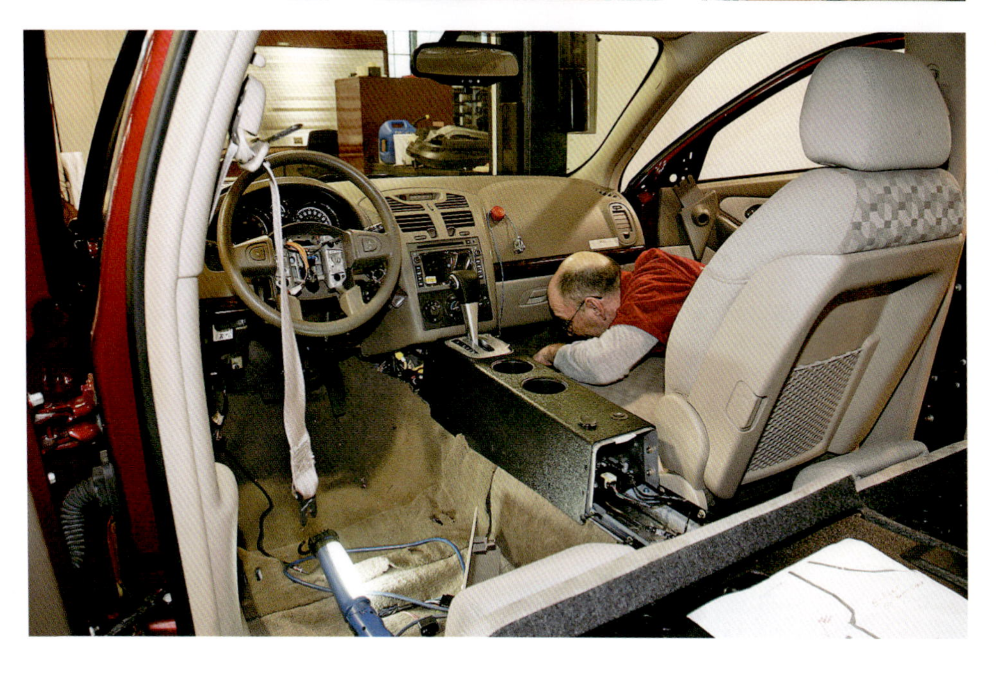

Inside the car, there are connections to be made and parts to be changed to transform the Malibu into a Volt prototype. *GM/Will Handzel*

Frank Weber is in the passenger's seat as GM vice chairman Bob Lutz gets ready for his first test drive in a MaliVolt mule. *GM/Will Handzel*

Top: A pair of Chevrolet Volt "mule" prototypes make their way around the test track at the GM Proving Grounds in Milford, Michigan, in mid-November 2008. *GM/Will Handzel*

Bottom: Jon Lauckner is behind the wheel of the other Volt prototype, maneuvering around the proving grounds on a snowy November day. *GM/Will Handzel*

Opposite: A lift gets ready to raise the chassis, so the battery pack and generator engine can be installed. *GM/Will Handzel*

From left, chief engineer Andrew Farah and hybrid battery engineer Larry Nitz explain the Chevrolet Volt battery-charging process to Environmental Protection Agency (EPA) directors Margo Oge and Chet France at the 2010 North American International Auto Show. *GM*

02
02
CAUTION
CAUTION
TRIP / FALL
PLEASE DO NOT STAND ON
BODY CARRIER

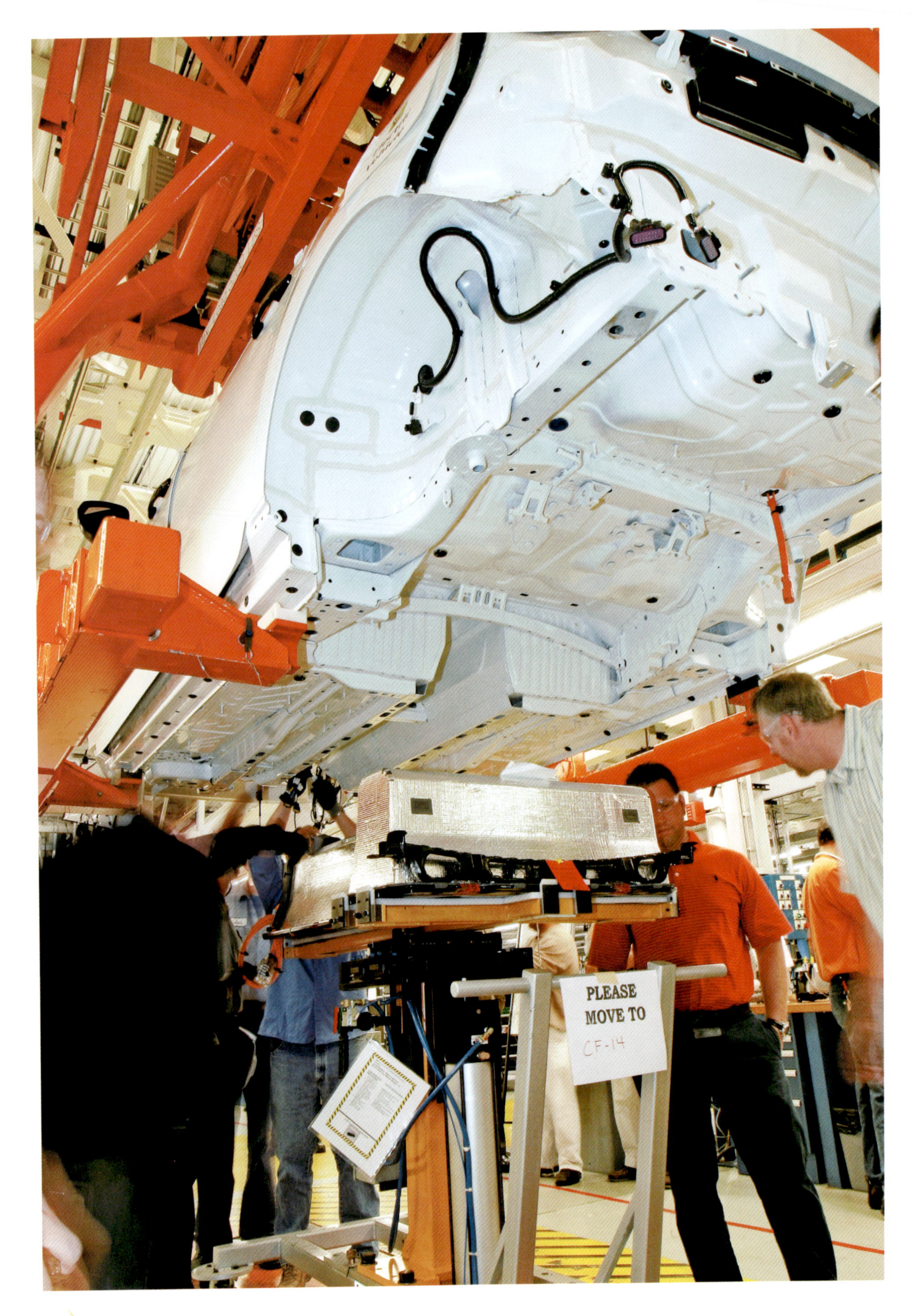
PLEASE
MOVE TO
CF-14

An electric-powered vehicle includes a lot of electrical wiring. *GM/Will Handzel*

Angles and radii are noted in detail prototype components. *GM/Will Handzel*

The body shell rides on a moving platform as it works its away down the assembly line in the IVER shop. *GM/Will Handzel*

The first Volt IVER moves through the shop. *GM/Will Handzel*

Opposite: The battery pack is installed beneath the Volt IVER's body shell. *GM/Will Handzel*

Various grilles, spoilers, and other parts undergo wind tunnel testing to verify the details of their aerodynamic contributions to reducing drag and providing airflow to cool internal components. *GM/Will Handzel*

Engineers and designers gather inside the General Motors wind tunnel in preparation for an aerodynamic test of the full-scale clay model of the Chevrolet Volt. *GM/Will Handzel*

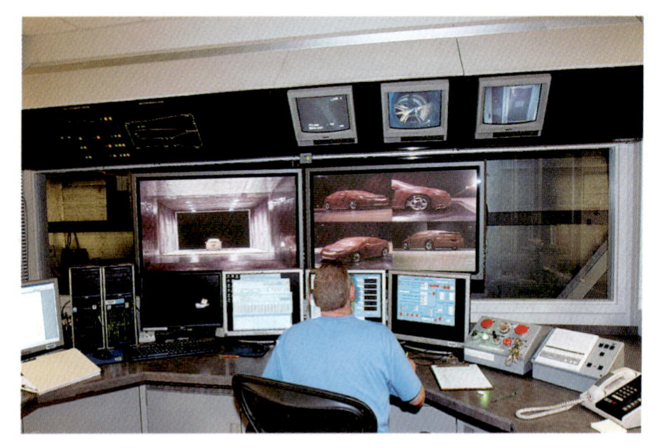

Video monitors and computer screens tracking various aerodynamic readings are studied carefully during wind tunnel tests. *GM/Will Handzel*

The devil truly is in the details. Here, engineers and designers gather around the car's tail to make sure every detail is both attractive and aerodynamically efficient. *GM/Will Handzel*

Opposite: The clay model can be modified right in the tunnel. Note the clay shavings that have fallen to the floor as the car's sculpture is changed. *GM/Will Handzel*

E-FLEX

CHAPTER 6

TUNNEL VISION

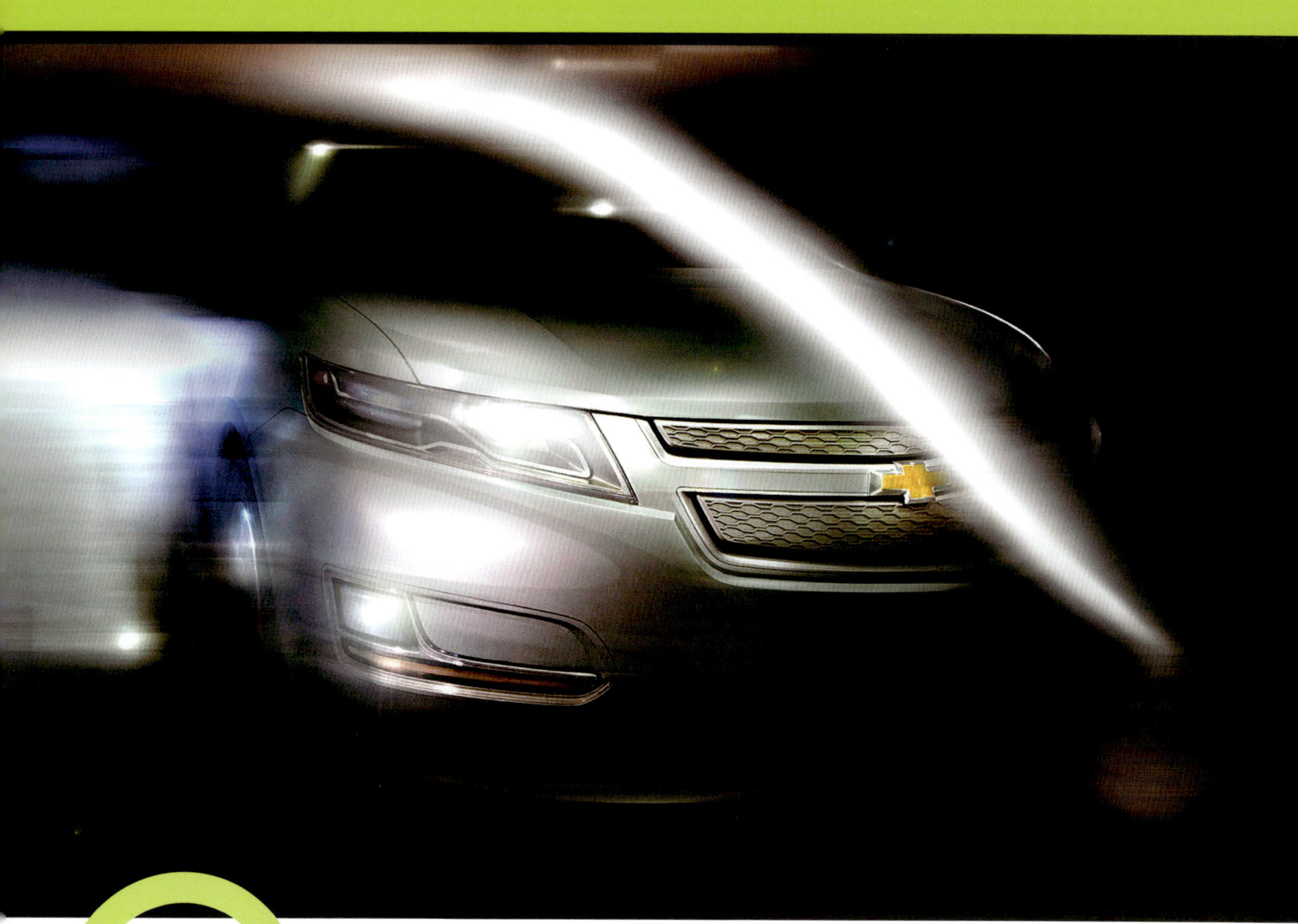

THE VOLT CONCEPT CAR WAS A GORGEOUS EXAMPLE OF AUTOMOTIVE SCULPTURE. DESIGN DIRECTOR BOB BONIFACE LIKENED IT TO AN "ELECTRIC PONY CAR." BUT WHEN IT CAME TIME TO TRANSFER THE VOLT FROM CONCEPT TO PRODUCTION VEHICLE, PHYSICS GOT IN THE WAY.

"PHYSICS DOESN'T KNOW WHAT YOU'RE WORKING ON. IT DOESN'T KNOW THAT IT'S AN ELECTRIC VEHICLE," SAYS NINA TORTOSA, THE GENERAL MOTORS AERODYNAMICS ENGINEER WHO OVERSAW WIND-TUNNEL TESTING THAT HELPED GIVE THE PRODUCTION VOLT THE OVERALL SHAPE AND EXTERIOR DETAILS NEEDED FOR IT OR ANY CAR OF THAT SIZE AND WEIGHT TO BE ABLE TO ACHIEVE UP TO 40 MILES OF TRAVEL ON BATTERY POWER.

WITH THE CONCEPT CAR, "WHAT WE WERE REALLY TRYING TO DO WAS TO BREAK FROM THE MINDSET THAT SAID EFFICIENT VEHICLES HAVE TO BE STODGY," BONIFACE EXPLAINS.

Above: For the Chevrolet Volt, beauty was not only in the eye of the beholder but in the sculptural efficiency developed in the General Motors wind tunnel. *GM*

But for the production version, fiscal and physical realities set in.

"We knew we weren't going to create a unique architecture for this car," he says. "We knew that the cost of the battery and the cost of all the propulsion system was going to be high." To keep those costs reasonable for the company and for customers, the production version of the Volt was based on GM's global compact vehicle architecture (which also was being used as the structural foundation beneath the Opel Astra in Europe and the Daewoo Lacetti in Asia and was being launched in North America in the form of the Chevrolet Cruze).

Among other things, that decision pulled the front wheels back from the corners of the concept car to a more conventional location, enabling the use of suspension and braking systems and a motor compartment that already had been validated in crash tests to meet safety regulations. The change also reduced the Volt's turning circle and enhanced its maneuverability.

"But we knew we were going to have to walk away from the proportions of the original Volt [concept]," Boniface says.

GM's global vice president for design, Ed Welburn, asked Boniface if he wanted to remain in the advanced design studio or to stay with the Volt through its production design.

"I jumped at the chance because this is an once-in-a-lifetime opportunity," Boniface said of the car's importance not only to General Motors but to the electrification of the automobile.

Most of Boniface's team who had worked on the Volt concept stayed with the car as well, though Jelani Aliyu, who had done the exterior design of the concept, opted to remain in the advanced studio. Young-Sun Kim, a native of South Korea and, like Aliyu, a graduate of the Center for Creative Studies, became exterior design manager. Kim had been design manager for the fuel-cell-powered GM Sequel concept vehicle (and for the drivable Chevrolet version of that car). He would serve as design manager for both the Volt concept and production vehicles and was, according to Boniface, "one of the unsung heroes of the program."

"We took the best theme elements from the show car, specifically the shape of the daylight opening, the shape of the tail lamps, the gesture of the roofline, the body side formations, the sweep line coming off the front wheel tracking toward the rear—all of that surface treatment was directly lifted from the show car," Boniface says. "And the rears of the cars are virtual clones of one another. They're very, very similar in proportion and how the pillar relates to the wheels.

"We had to add a few aerodynamic devices to the back of the car, obviously, with the spoiler height and those blade forms we put on the trailing edge of the fenders, all in an effort to get the drag number down."

Boniface says the two biggest challenges in doing the production version of the Volt were packaging and aerodynamics.

"Packaging [driven primarily by the powertrain and the dimensions of the battery pack] is what drove the proportional changes more than anything," he says. "But aero was the biggest challenge because you can't lie to the wind. That fan is going to tell you if you succeeded or not."

Below and following page: Taking the Chevrolet Volt electric vehicle with extended-range capability from a concept car to a production model necessitated design changes both inside and out. These designers' sketches show some of the evolution of the exterior design, which was needed because the production car would need to be built on GM's global compact car platform rather than a one-of-a-kind concept car chassis. *GM*

"You do certain things to the car and the number doesn't go down. . . . Come on! . . . Come on! . . . You start flogging the car. But if you work with a talented aerodynamicist like Nina, she makes recommendations: 'If you want to get the number down, you should do this . . . ' There were certain things when I would say, 'OK, I'll do that,' but there were other things where I would say, 'This is absolutely a nonstarter.'

"For instance, one of the quickest ways to drop the drag coefficient of a car is to put a long tail on it. However, the car wouldn't have looked good. We said this car is going to be a compact; it's going to have tidy dimensions. It's going to look efficient. And putting a big long tail on the car is not going to do it. 'OK,' she said, 'then you're going to have to do some type of device back here to trip the air,' and that's why you see that crease that runs down the edge of the rear quarter panel. To me, it actually enhanced the aesthetic of the car. It looks functional, like it's there for a reason."

Which it is, because the laws of physics are, indeed, laws that cannot be violated.

In the design studio, they tease that the Volt was actually designed across the street in the General Motors wind tunnel. Nina Tortosa laughs at the suggestion that she had her way with the car.

"I just report how physics works," she says, admitting that because of the way physics works and the requirement that the car be able to travel up to 40 miles on battery power, the Volt spent more than 500 hours in the wind tunnel (200 hours is the target for a typical production car program) and that nearly 1,100 changes—again much higher than the typical number of alterations—were made to the details of the design.

Tortosa was born in Spain but in 1982 was living in Denmark, where her father was a chemical engineer.

"I was in sixth grade," Tortosa says as she remembers watching the liftoff of the first shuttle mission on television. "I wanted to be an astronaut."

Her family moved to the United States, to Minnesota's Twin Cities, while Tortosa was still in grade school, and she studied science and math and planned to become an engineer and

This is the clay model of the Chevrolet Volt as it stood in late June 2007. Note that the model shows two design themes, one on its left side and the other on its right. *GM*

Design managers Bob Boniface (left), Young-Sun Kim (center), and Mike Simcoe study details of a Chevrolet Volt model in August 2007. Like many clay models, this full-scale car includes two designs, one on each side of the car, and the clay has been covered with a plastic fabric that more closely mimics how the car will appear with a sheetmetal skin. *GM*

This clay model, photographed in September 2007, has its left side covered in fabric, but its right side still is in clay. *GM*

astronaut. After high school, she enrolled at the University of Minnesota, where she earned a master's degree in aerospace engineering and mechanics, and in 2000 she went to work dealing not with outer space but ground-bound aerodynamics at General Motors.

She recalls that she began working on the Volt even before it was the Volt. She started her career at GM doing CSV analysis—computational flow dynamics—basically a computer-based version of wind-tunnel testing, and she ran numbers on the early math models of a proposed electric-powered concept vehicle that at first was referred to as the iCar and ended up being called the Volt.

Nina Tortosa handles the smoke wand, one of the tools used to see how efficiently air flows over, around, and even beneath the Chevrolet Volt in the General Motors wind tunnel. *GM*

GM executives—that's Jon Lauckner at the far left, Bob Lutz with his left hand on his hip, and Ed Welburn next to Lutz—join design staff leaders in examining a Volt model in September 2007. *GM*

Then, to turn the concept into a production vehicle, she spent the equivalent of nearly 60 work days in the wind tunnel with the car, working at first with one-third-scale models, then doing the fine detail work on a full-scale version.

"To get the forty miles of electric vehicle range we needed, that early concept wasn't going to cut it," she says. "We started playing around with the clay. It's an iterative process where the designers will come to the wind tunnel with me and I'll make suggestions: . . . We usually start at the front of the vehicle and work our way back. They'll say, 'I don't like this, I don't like that,' and I'll say, 'Think about this . . . ' You get together and try stuff, and they take it back over to the design center and they get the people in charge to look at it and they say yes, no, or try something else.

"It was a really good team. Bob Boniface was very supportive of aero, and had he not been we wouldn't have gotten as far as we did."

Changes are made first using small-scale models, then a full-scale model is created to get a true gauge on aesthetics, and then the scale model goes back into the wind tunnel to be graded on aero effectiveness.

"We test, we modify it further, and start the process over again," Tortosa says, adding that in addition to the general aerodynamic challenge, there also are issues with making sure there's proper airflow for cooling not only in the typical engine but, in the Volt's case, the other and even more critical powertrain components.

After final tweaking, the full-scale model of the Volt is rolled out in September 2008 for final approval. *GM*

"The conversations don't get violent," she insists. Though she admits, "There are one or two where the designers and I don't agree and they want to do something and it's not going to work, and [I'll say,] 'I really think you need to look at this,' and they don't want to, but in the end we work to do what's best for the car, so we try their ideas and we try my ideas. You have to try it. If you don't try, you don't know.

"In the end, we found a good balance where they kept a lot of the original design cues and got the fuel economy and EV range."

Though the back end of the production Volt may look like the back end of the concept, Tortosa says that's where the biggest changes were made in the tunnel to make sure the airflow was smooth rather than creating range-ruining turbulence. There also was significant aerodynamic work done on the underside of the vehicle to help it slice through the air without wasting power.

Changes, Tortosa says, often are measured in saving the equivalent of a 10th of a mile per gallon, but when you add up such small amounts you can get surprisingly impressive benefits. Boniface notes that redoing the barely perceptible to the eye five-millimeter radius of the Volt's rear spoiler produced an aerodynamic improvement that added a quarter-mile to the car's electric range.

"You'll find things like that all over the car," he says.

The one-third-scale models are used until the car is within 2 percent of its aerodynamic requirements, Tortosa says. Then, only full-scale models are used.

"We like to do the big clay changes in third-scale because for every pound of clay you're moving in third-scale, you're moving twenty-seven pounds in full scale," she says.

"In full scale, you can do a lot more detailed A-pillar shapes, the mirror development needs to be done in full size, and underbody panels and air dams. We can work on the details, the radiuses and much smaller changes."

Take rearview mirrors, for example.

"In mirrors, you're looking at one- to five-mill radiuses on the edges. When you go to third-scale, you have to divide that by three, so if you're doing

Young-Sun Kim was the manager for the exterior design of the Chevrolet Volt. *GM*

Gary Ruiz was an exterior designer for the Chevrolet Volt. *GM*

In Ho Song was an exterior designer for the Chevrolet Volt. *GM*

Though tuned in the wind tunnel, the 2011 Chevrolet Volt carried themes from the concept car into the production version. *GM*

a one-millimeter edge on something, you're going to a third of a millimeter, and that's the thickness of the tape that you're using to tape the mirrors onto the clay model."

How much of an impact does something so small have on overall drag?

Reduce the drag on a mirror and, since there are two of them on the car, you get a double benefit. Plus, mirrors can create a lot of unwanted wind noise.

"For the Volt, we ended up going with stalk-mounted mirrors because they're further away from the side glass, so they interact less and, generally, the stalk-mounted mirrors are lower-drag mirrors," Tortosa explains. "Ours are on par with the Corvette C6 mirrors."

And, she says, "It does add up. One thing here. One thing there. Each may not be a big deal, but when you add them all up you have quite a bit, and on the Volt you have both miles per gallon but also electric-only range, which is a different equation.

"Because it's an electric vehicle, there is a lot of mass in the battery storage system, so aerodynamics [to make up for the increased mass] had a bigger chunk of the pie than [they] do on most programs.

"Design is protective of their designs, as they should be, that's their job," Tortosa says. "But it's a team effort. There are things in the tunnel that worked for drag that I didn't like the way it looked. I like cool cars just as much as anyone else.

"But my job isn't just to make it look good, it's to show them what reduces drag."

"You have tradeoffs for cost, for weight, for investment, for aerodynamic performance," Boniface adds. "But on this car, since you're already strapped with huge cost in the battery, huge investment in the powertrain, huge aerodynamic challenges, each one of those standard production car issues is magnified. That's why I'm so happy about how the car ended up. We were able to meet all our targets—and there were some lofty goals."

CHAPTER 7

SCREEN PLAY: THE VOLT INTERIOR STORY

The printed press kit distributed at the unveiling of the Chevrolet Volt concept devotes two of its 26 pages to three photographs of the car's interior and one that shows interior design manager Wade Bryant and lead interior designer Therese Tant. It notes that the styling theme for the interior originated in General Motors' British studio and that "near-term technologies and innovative materials combine with ingenious use of ambient light for an interior environment that's light, airy, and thoughtful."

A compact disc that accompanied the press kit explains that the interior theme is "city lights":

The Volt's roof, side glass, and beltline are constructed of a transparent, glazed polycarbonate material that delivers the scratch resistance and gloss surface appearance of glass, combined with the formability of a plastic composite. As a result, the Volt provides the driver and occupants with exceptional visibility, enabling a "city lights" theme in which the outside world passes through to the interior of the vehicle. Also contributing to the visibility is a shouldered, tinted side glass—constructed of the same polycarbonate material—that enables a dual beltline.

Bryant is quoted on the disc explaining that the interior environment was designed to appeal to an urban dweller who wants a smart, daily-use vehicle:

On the interior of the Volt, you'll find technologies, materials, and an environment that enable the car to help make life simpler for a person who's environmentally conscious and leads a city-centered lifestyle. It's ergonomically correct, provides connectivity to the world, and demonstrates smart responsibility through the use of lightweight, renewable materials.

Bryant notes how the exterior and interior design teams worked together to give the Volt a unified appearance. For example, the door hinge, which is functional both on the interior and exterior, becomes the pull handle for closing the door inside the vehicle.

The dual-mode instrumentation was devised to reflect the potential provided by the powertrain's plug-in capability.

"The powertrain technology is the key feature, so we wanted to make sure the interior communicated that, and the driver would have a sophisticated, fun, and useful interaction with the electric-drive system," Bryant adds, referencing the way the information is displayed on the dashboard.

The primary level of information is configured similarly to a conventional instrument cluster and provides traditional data in analog but also has three-dimensional displays employing light-emitting diodes. These include gauges for the fuel level and battery level, the transmission indicator, the speedometer, and the odometer.

The second level of information, shown on a screen near the base of the windshield, displays holographic-style color-animated data related to the Volt's advanced propulsion system. This system uses invisible, fluorescent inks printed on a transparent screen. When illuminated by an ultraviolet laser projector located behind the instrument cluster, the inks provide four-color illumination and animation.

Another interior feature is a compression-molded foam with a textile-patterned surface layer that was applied on the lower instrument panel, lower door trim panels, and rear quarter trim areas. This material provides soft, tactile, low-gloss surfaces that look hand-crafted and tailored to the Volt.

The material was inspired by that used in luggage design, and that theme carried over in the use of zippered access to the glove box and other storage compartments. Those zippers are transparent—and for a reason:

"All the storage areas are lighted internally," Bryant explains, "and the light escapes through the clear zippers, so you'll always be able to find your storage at night."

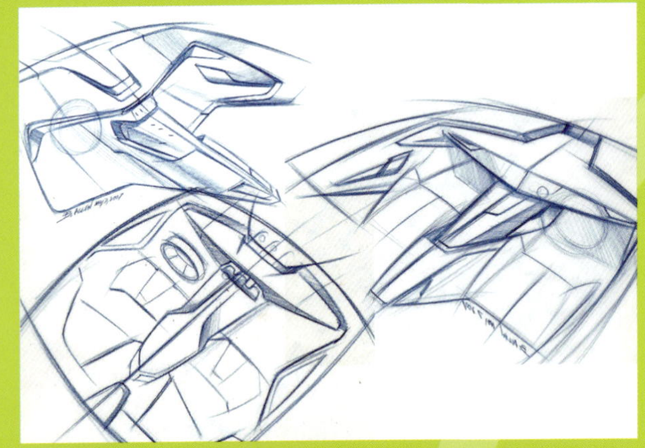

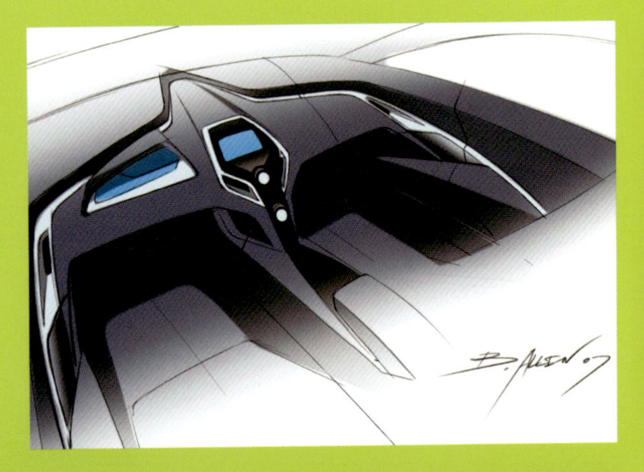

wip

Each of these clusters shows interior design proposals from one of the General Motors designers. Each designer works out his or her own themes. One designer's theme may be selected for further development, often incorporating elements from other designers' sketches as the studio seeks the optimum interior for the vehicle. *GM*

VOLT
J. GASPAROTTO 07

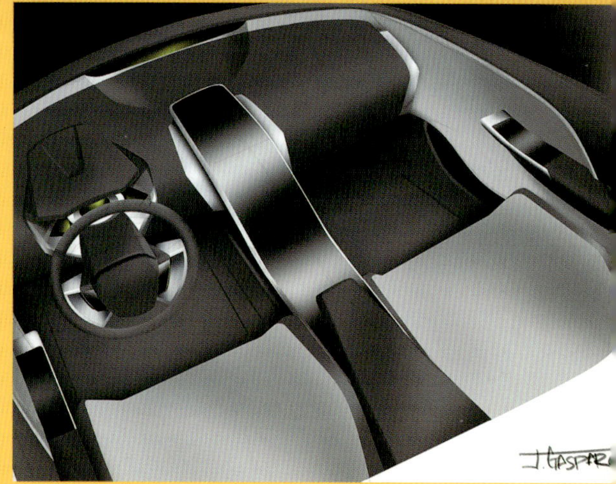

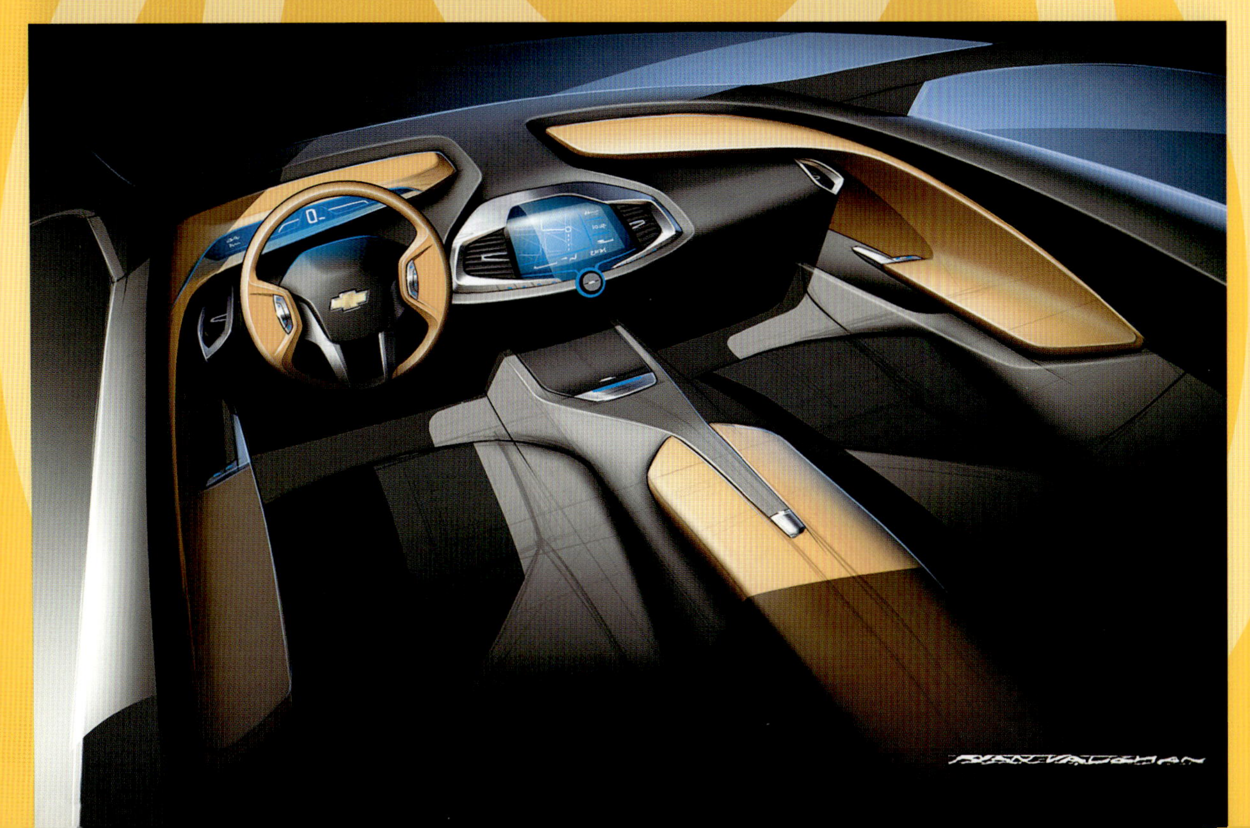

The GE/GM technology also allows the use of conductive ink controls printed on the interior surface of the roof. The inks illuminate at night and provide "touch access" to turn on the interior lights or to communicate with OnStar services.

Seats and other interior components employ thin but structural surfaces and appear cantilevered in space because of the use of molded plastic panels provided by GE. These light, strong, affordable, and recyclable panels are sheathed in reconstructed scrap leather that provides a soft and hand-crafted surface.

Bryant also notes in the press kit compact disc that all of the technological innovation showcased in the concept car's interior "is definitely based in reality. All the things you see on the interior are within reach in the next few years."

Theresa Priebe was the lead interior designer for the Chevrolet Volt. *GM*

Tim Greig was manager for the interior design of the Chevrolet Volt. *GM*

The challenge for Tim Greig and the designers working in his studio was multi-faceted. For one thing, they didn't have a few years to wait for such technologies to be developed for practical application. For another, they had to create an interior not for a one-off concept car but for a production car that would meet all standards and perhaps even exceed expectations for safety, durability, aesthetic appeal, and affordability.

"There were several challenges," says Greig, whose father, Richard, worked for General Motors as an assembly plant engineer. Richard was a mechanical engineer who packaged General Motors assembly plants, deciding where components, such as ovens and booths and conveyors, should go to enhance production while keeping plant footprints as compact as possible. His son faced similar challenges in leading the interior design of the Volt.

Richard Greig was assigned to GM's Brazilian division from the time Tim was nine years old until he was a senior in high school. At one point during that period, Tim was sick and bedridden for several days. To pass the time, he traced images of cars from automotive magazines and added his own modifications. His father took some of the sketches to work and showed them to GM designers, who invited Tim to visit the design studio.

"I was maybe thirteen or forteen at the time," Tim Greig recalls. "They invited me in for a personal tour. They were doing a facelift for a pickup truck, and there was all this artwork up on the wall, and it just grabbed me. That was it. My decision was made."

After high school, Tim enrolled at the Center for Creative Studies in Detroit and joined the GM design staff after his graduation. At one point he was assigned to the Brazilian studio for six months, now working with some of the same people who had given him his private tour some 15 years earlier.

At GM, the younger Greig spent much of his time in Pontiac studios, where he helped do the interiors of both concept and production cars, including the Banshee and Pursuit concepts and the Grand Prix. Greig also spent time

Once the sketches are developed, an "interior buck" is created so designers and executives can see those images translated as full-scale and three-dimensional models. The buck is made from a wood and metal frame, covered with clay sculpted into forms proposed for production. *GM*

with GM in Germany, which is where he began work on what would be the exterior and interior of the Chevrolet Cobalt. He did the interior for the Chevrolet HHR, and then he worked on a couple of projects that were canceled before reaching production.

He was brought in to work on the Volt as the concept was being completed and to lead the development of the interior design for the production version of the car.

"Certainly from a packaging standpoint, the battery pack was a curveball to us," Greig says, noting that the large, T-shaped power source intruding into the space usually reserved for occupants had the interior designers playing "a very different game."

Another challenge: "Shrinking the exterior down, so the frontal area is as small as possible for aero, obviously eats into our real estate [inside the car]." Complicating this challenge was that battery pack "pushing up and out," Greig adds. As a result, the four occupants had to be moved 12.5 mills—nearly half an inch—outboard from their usual seating positions with no allowance to make the car any wider. Such a shift—it's only half an inch—may not sound like a big deal, but consider that there can be no compromise in protecting those occupants in a side impact, and thus seats, door structures, and even details of the car's B-pillars have to be redesigned, re-engineered, tested, crashed, and validated.

Yet another challenge: "Perception," Greig says. "It's an electric car. It has an onboard generator. So there's just a ton of technology and information that you could convey to the customer. But to a lot of people, new technology can be scary."

That's why the information must be presented in an appealing way. "It had to be warm and inviting and encourage you to actually want to get in instead of being threatened by it because 'it's too much technology for me,'" Greig says.

One important enabler was to either discover or even develop what Greig and his designers called "an electric aesthetic."

"We wanted to own it, to plant the flag and say, 'We got it!'

"One of our biggest challenges, aside from the packaging, was [answering the question], 'What is this electric aesthetic?'"

Oh, yes, the interior also had to reflect the Chevrolet brand character, itself an evolving aesthetic significantly enhanced with the launch of vehicles like the latest iterations of the Malibu, Equinox, and Camaro.

As far as an electric aesthetic, and especially the "warm tech" appeal Greig and his studio wanted, what they discovered was, well, there really wasn't one. The closest they could find was the clean and modern design of Apple's iPod and iPhone, not only in the ease-of-use, hold-it-in-your-hand design, but because of the use of high-value materials—real glass and real metal.

"Some people are going to buy the car because it's the coolest thing on the block."

—TIM GREIG

"The shape is friendly," Greig and his designers note. "You don't see sharp edges. And the weight of it. And the interface. Things just don't appear; they kind of grow to the surface or they slide to the side, so you have this very organic interface to the customer."

It quickly became apparent that the key to the electric aesthetic the designers wanted for the Volt's interior was more than materials; it was in the controls and information displays.

"The layout of those controls and the design of those controls needed to be absolutely intuitive," Greig says. "In fact," he adds, "just looking at them needed to be intuitive."

It became apparent that the controls and displays would provide the warm but high-tech signals while the rest of the interior would provide the Chevrolet brand character with an inviting look that seemed to pull you right into your seat.

"We wanted very clean surfaces, very new materials, very high-tech-looking materials," Greig says. "And we wanted the user interface to be absolutely flawless, in fact, engaging and compelling."

It also became obvious that with so much information available, "a conventional display would not work," Greig says. To provide all the available information, and in an intuitive way not only for the driver but also for the passengers, he says, you'd need the equivalent of two laptop screens. However, you'd need those screens configured so people wouldn't be overwhelmed.

"Some people are going to buy the car because it's the coolest thing on the block," Greig explains. "Others are techno enthusiasts. Some want to get the country off petroleum. Some are going to buy just because it's a smart purchase—I save thousands of dollars by not burning gasoline.

"The customers who are concerned about just saving money may not need all the information the techno enthusiasts want."

"Instead of one solution for all, we have different modes you can select, so if you don't want all that information, you can press a simplify button and it will give you very basic information—your range, how much energy you have, your speed," he adds. "But the techno geek can see it all." Even download it to a laptop or smart phone to share with others of his or her ilk.

To design the controls and displays, GM Design turned to its own techno advocates, the HMI—Human Machine Interface—group led by Stuart Norris, an industrial design graduate of Coventry University in England who, he admits, is a real geek who enjoys the annual Consumer Electronic Show more than any auto show.

Norris worked at Jaguar before joining General Motors, where he did midsize truck interiors before focusing on HMI, which took a big leap at GM with Norris' work on the 2008 Hummer HX concept vehicle (it had three display screens arrayed across its dash).

"The Volt is a revolution. It's a different way. This is entirely different at every level. There's no traditional engine noise. There's no engine revs going up [and down on a tachometer]," Norris says, adding that many of the usual gauge displays are nonfactors on the Volt.

"It's also about trying to appeal to a different kind of customer. This kind of customer is not going to be looking for a red line on their tach and whether they have a one-hundred-forty- or a one-hundred-eighty-mile-per-hour limit on their speedometer. This customer is looking for a different kind of experience."

Volt buyers, he says, are likely to be very tech savvy, very comfortable with Wii and Xboxes, Twitter and iPhones.

Almost immediately, Norris says, design moved away from thoughts of simply using the display screens to show traditional gauge images. Instead, the displays are seen as visual space, space with depth and perspective.

"It is designed to accommodate the kind of powertrain that this vehicle has . . . with beautiful, juicy renderings," he says.

In simplified terms, the screen in front of the driver shows the vehicle's speed, position of the gear selector, odometer, and anticipated range on battery power. When the vehicle transitions to range-extending mode, the display changes to show the fuel tank and anticipated overall range.

Multiple components may be built so decisions can be made about a vehicle's interior details. Note the differences in the three dashboard proposals. *GM*

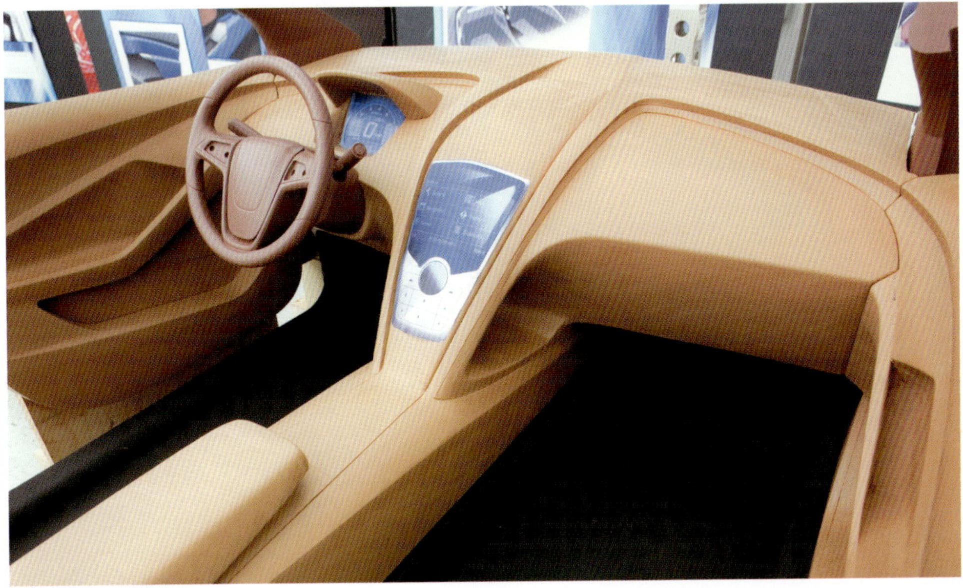

Mac Nolan and Erin Aube use tape and carving tools to create details in the clay model of a door panel and dashboard. *GM*

Even the seats are sculpted in clay. Richard Cruger sculpts the edge of a Volt seat. *GM*

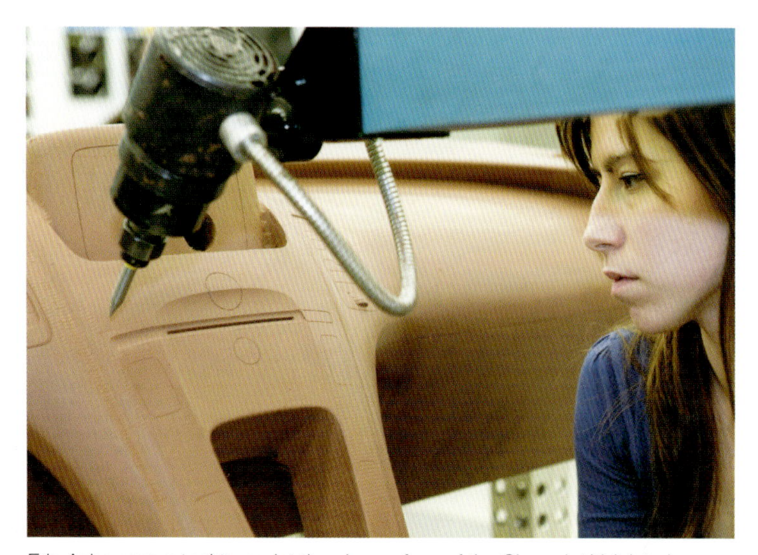

Erin Aube uses a tool to sculpt the clay surface of the Chevrolet Volt interior styling buck. *GM*

"There's a nice kind of transition when you change to extended-range mode," Norris says. "The whole layout of that graphic was designed with the idea that I have an electric vehicle that sometimes becomes an extended-range vehicle."

Drivers can opt for various depths of information on the screen in front of them. The screen on top of the center stack can be even more widely reconfigured to provide even more information, not only about the vehicle but its various systems—navigation, radio, air conditioning, and current and historic powertrain displays.

"It's really very much along the lines of where the iPhone is, where many other technologies are today," Norris says. "It's a software-based interface, so it's infinitely updatable. This is the last cluster you will ever need because you can continue to rewrite the software and reflash it on there if we decide we want to make this look a little different or to reconfigure it in a different way, or if there's more information we want to add. It's infinitely reconfigurable and updatable and upgradable."

Part of the design process involved research in the General Motors driving simulator to make sure the dazzling displays won't cause driver distraction.

"The great thing about reconfigurability is that you can do lots and you can do nothing at all," Norris says. "That option is there. We have a layout that's simplified and then an enhanced interface. The simplified interface tells you your extended range, your EV range, what your speed is, and what gear you're in. That's pretty minimal information."

Or, he says, you can opt to include everything from tire pressure information to gauges of your driving efficiency.

"The approach we took between the [driver] cluster and the center stack is with the cluster we wanted the information that can affect your driving now," Norris says. "There's an efficiency gauge that shows I'm braking too aggressively [and thus not getting full feedback from the regenerating braking system that feeds power back to the batteries], or that I'm accelerating too aggressively [and thus wasting power].

"Then, in the center stack, we have lifetime fuel economy and fuel economy over a given trip, and it also gives you information about how efficiently my HVAC [heating, ventilation, and air conditioning] is operating."

"HVAC," he notes, "is one of the single biggest factors in terms of affecting electric driving range."

"The center stack stuff is that next level," he continues. "We have a tips screen. You can go into a green leaf menu and look at your efficiency over time. Or you can watch how the power flows [from generator to battery to wheels] and how your regenerative braking is working. And there's a charge scheduling layout. I can go in there when the vehicle's in Park. I can set my vehicle to start charging at seven o'clock in the morning and finish at eight o'clock at night. I can plug in my local electricity rates so it shows if I'm using the most efficient, cost-effective times on the charging schedule.

"On top of that, we have OnStar and some of our mobile apps stuff that allows an extra level of information to help you understand the efficiency of your driving."

The center stack has a touch screen, though there's also a redundant rotary-style controller. The driver's screen has a controller mounted on the dashboard to the left side of the steering wheel. "You don't want people putting an arm through the steering wheel [circumference]," Norris notes.

"The jump we've made with the Volt displays is very, very significant," Norris adds. "I think it will flow to other GM vehicles, because once you've had a taste of this stuff . . . "

Norris says his studio includes people with computer gaming and movie studio animation skills. "We have stunning stuff on the Volt," he adds. "The two screens are electronically tagged to each other. Under normal circumstances a cluster lives here and a center stack lives here and never the twain shall meet. On the Volt, we got them to talk to each other so we could do animations between the two. There's a start-up animation where there are elements that come from the center stack screen and move into

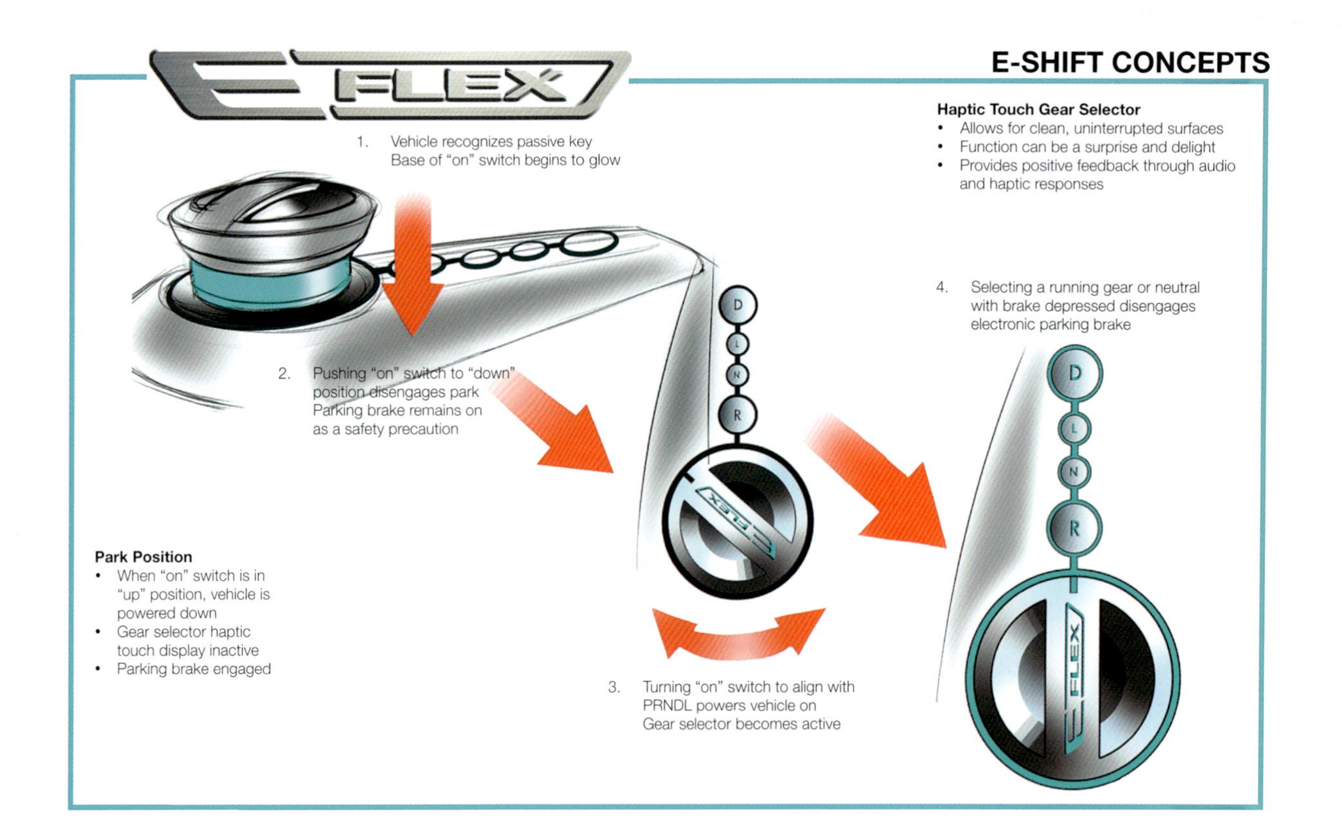
E-FLEX
E-SHIFT CONCEPTS
Haptic Touch Gear Selector
• Allows for clean, uninterrupted surfaces
• Function can be a surprise and delight
• Provides positive feedback through audio and haptic responses
1. Vehicle recognizes passive key
Base of "on" switch begins to glow
2. Pushing "on" switch to "down" position disengages park
Parking brake remains on as a safety precaution
3. Turning "on" switch to align with PRNDL powers vehicle on
Gear selector becomes active
4. Selecting a running gear or neutral with brake depressed disengages electronic parking brake
Park Position
• When "on" switch is in "up" position, vehicle is powered down
• Gear selector haptic touch display inactive
• Parking brake engaged
D
L
N
R

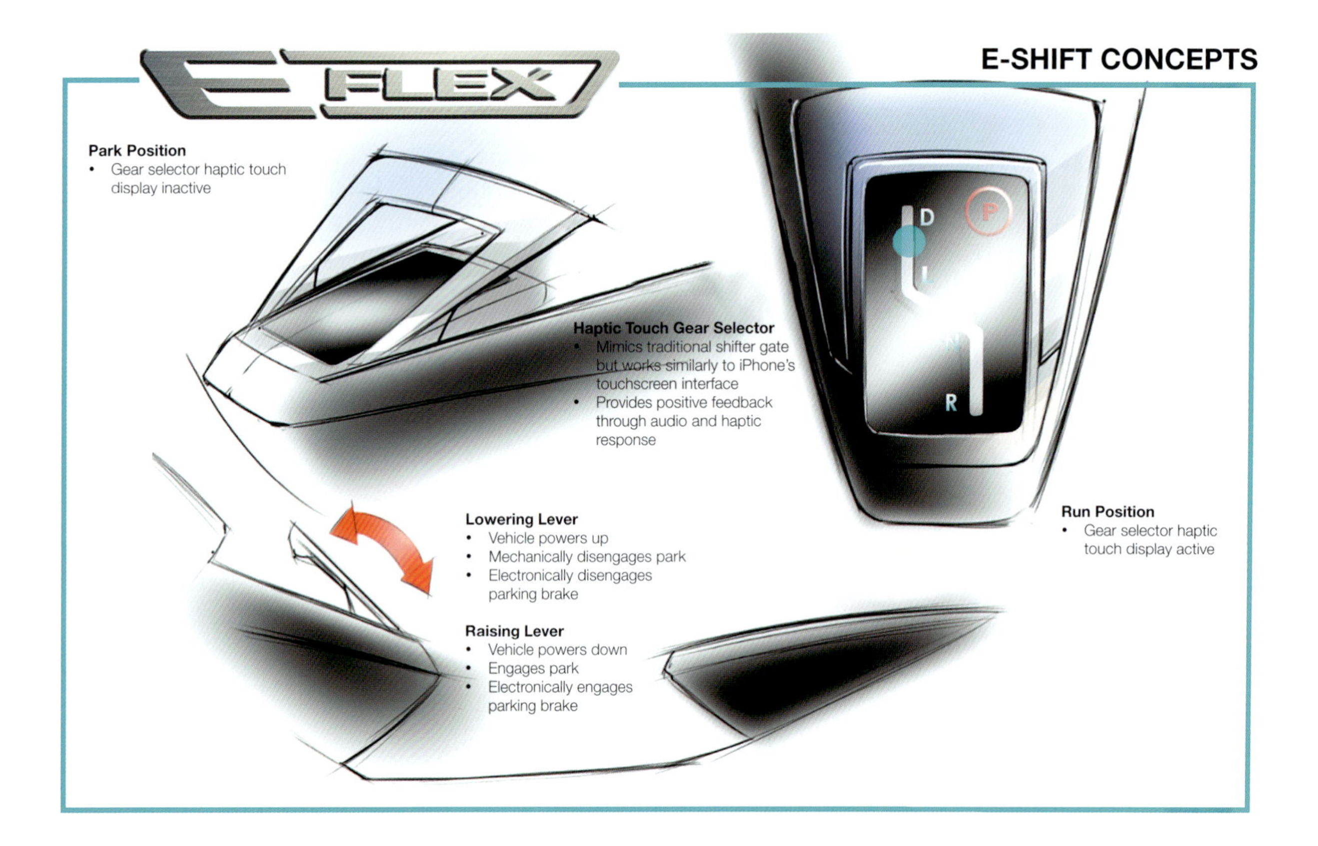
E-FLEX
E-SHIFT CONCEPTS
Park Position
• Gear selector haptic touch display inactive
Haptic Touch Gear Selector
• Mimics traditional shifter gate but works similarly to iPhone's touchscreen interface
• Provides positive feedback through audio and haptic response
Lowering Lever
• Vehicle powers up
• Mechanically disengages park
• Electronically disengages parking brake
Raising Lever
• Vehicle powers down
• Engages park
• Electronically engages parking brake
Run Position
• Gear selector haptic touch display active
D
P
R

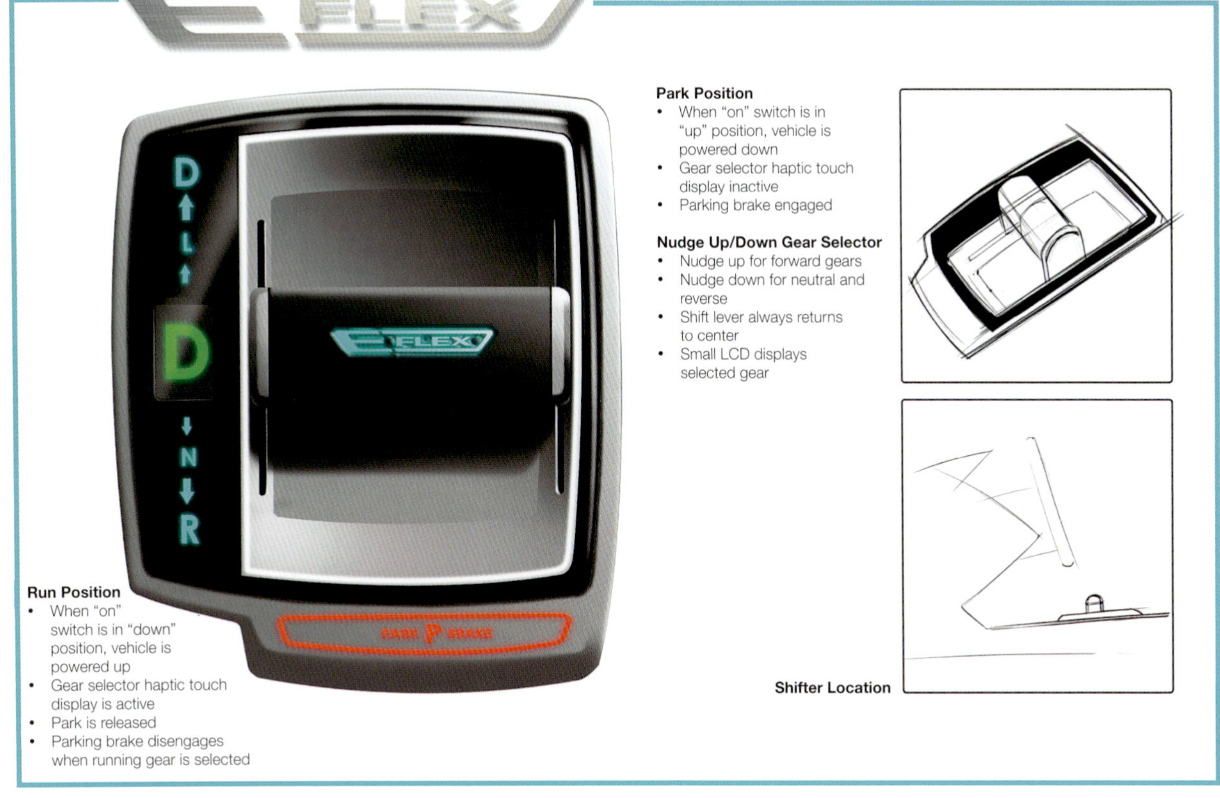

the cluster. I think [video images of] that stuff will turn up on YouTube when the car gets launched."

CONSOLE-MOUNTED E-SHIFT
FLEX
D
L
D
N
R
PARK P BRAKE
Park Position
• When "on" switch is in "up" position, vehicle is powered down
• Gear selector haptic touch display inactive
• Parking brake engaged
Nudge Up/Down Gear Selector
• Nudge up for forward gears
• Nudge down for neutral and reverse
• Shift lever always returns to center
• Small LCD displays selected gear
Run Position
• When "on" switch is in "down" position, vehicle is powered up
• Gear selector haptic touch display is active
• Park is released
• Parking brake disengages when running gear is selected
Shifter Location

Norris adds that even with its Chevrolet badge, "you could easily say the Volt is a luxury vehicle, but it's not luxury in the traditional execution of luxury. It's not about wood. It's not about leather. It's not about chrome. When we were first talking about reconfigurable clusters, we were talking about them for Cadillacs."

The luxury, he went on, is in the "gorgeous, jewel-like" quality of the display. "It's only a seven-inch display. It's not like a fifteen-inch MacBook Pro. But there's something about the delicacy and beauty and the precision of how each little pixel is rendered on that display that transcends traditional luxury and really talks about a new language of technology and luxury."

But there's much more to the Volt interior than the two display screens or the amazing capacitive-touch controls that react to the electric current in your fingertips.

"Comfort was another area in which we were not going to compromise," Tim Greig says, listing the seating surfaces, surfaces where you might rest your arm or hand, even the storage compartments as areas where driver or passenger comfort were emphasized in the design process. In areas where, say, plastics were used, they were combined with what Greig calls "unconventional approaches," such as the use of high-gloss acrylics in the interior door panels, which can be ordered with graphics that emphasize the electric aesthetic.

As with any vehicle program, there was give and take. Designers tried everything they could do to figure out a way to fit a third person into the back seat, but in the end there wasn't even room for an integrated child's seat atop the middle section of the battery pack. But while they couldn't fit a third person, they did take advantage of the available real estate for cup holders, storage, and an auxiliary power outlet that will make the back seat more enjoyable for those sitting there.

Everything in a car has to be designed and engineered. Here are three options for the gear-shift selector system. *GM*

Above: Eventually, the clay model bucks are covered with materials that show the proposed fit and finishes for production. Here are two themes for the Volt interior that went forward until one was selected. *GM*

Opposite: The final buck for the 2011 Chevrolet Volt interior emerges. *GM*

They wanted to use renewable or recycled materials where possible but couldn't find enough of automotive grade and durability. However, they were able to use soy-based foam inside the Volt's four bucket seats. Again, there was a bonus: the soy foam was lighter in weight.

Greig notes that even the time constraints his team faced ended up working to the team's advantage and, he trusts, the customers' advantage: "We did this interior in roughly half the time we would normally do an interior," he explains, "but that worked to our advantage because there were fewer opportunities for people to say no."

Besides, he adds, "it was hard to argue with what we were doing to keep the interior warm and inviting. We had a pretty strong argument that this was a pretty rational design, compelling and highly emotional. It definitely looks high-tech but also 'gotta have' inviting."

An alternative solution was selected, and then it had to be modified once engineers began driving prototype vehicles. While elegant and beautiful, the shifter didn't leave enough clearance for the driver's fingers. *GM*

CHAPTER 8

SURGE PROTECTORS

A REVOLUTIONARY EXTENDED-RANGE ELECTRIC POWERTRAIN, AN EXTERIOR DESIGN SCULPTED INSIDE THE WIND TUNNEL, AND AN INTERIOR CREATED TO MAKE HIGH-TECH WARM AND INVITING WERE THREE MAJOR ELEMENTS IN THE DESIGN AND DEVELOPMENT OF THE CHEVROLET VOLT. BUT THERE WERE OTHER COMPONENTS TO THE CAR'S DEVELOPMENT THAT ALSO WERE CRUCIAL—FROM MAKING SURE THAT IT WOULD FIT INTO A CHEVROLET DEALER'S SHOWROOM TO MAKING SURE IT WOULD PROTECT ITS OCCUPANTS IN THE WORST-CASE SCENARIO OF A MISHAP ON THE HIGHWAY.

ALONG THE WAY, THERE ALSO WERE SOME PERSONNEL CHANGES. THE NEWCOMERS NOT ONLY KEPT THE VOLT ON THE RIGHT ROAD, BUT THEY BROUGHT NEW SKILL SETS THAT ACCELERATED THE CAR'S PROGRESS.

Above: A Chevrolet Volt race car? That was the idea behind the design of the Chaparral 3E, a concept for a Grand Prix–style race car with a Volt-style powertrain. *GM*

"I was the 'Chevrolet' person," said Cristi Landy, who was part of the Volt development team from the very beginning.

At many car companies, Landy's title would be product planner. However, at General Motors, her role is known as product manager, the representative from the Chevrolet marketing department with responsibility for knowing the voice of the customer and for making sure that voice is heard during the vehicle design and development process. Landy, educated as an electrical engineer, was a veteran at this exercise. Before taking on the Volt, she had been a product planner at Saturn and for GM's Lambda platform. She had also been a marketing manager for the newest version of the Chevrolet Corvette, the C6. Launched in 2005, it was widely acknowledged not only as America's sports car but a truly world-class supercar.

But not only is the Corvette a world-class supercar, it is the most affordable of such vehicles. Such affordability is a hallmark of Chevrolet.

"Chevrolet is the face of General Motors," Landy says, adding that the extended-range Volt also was designed to be a "solution for the masses" when it comes to electric vehicles. As a Chevrolet, the Volt needed to be affordable, though when cutting-edge technology is involved, affordability can become a relative term.

"It is affordable," she says, "though unfortunately, anything with new technology comes in at a higher price."

"Lithium-ion batteries are great, but not cheap," she adds, noting that two other electric vehicle brands that use li-ion battery systems—Tesla and Fisker—are much more expensive. Though Landy was part of the original iCar concept vehicle planning team as early as March 2006, it wasn't until after the Volt concept was unveiled at the Detroit show early in 2007 that serious research began with potential consumers. Focus groups incorporated owners of midsize cars, people who owned hybrids, electric vehicle owners—some who had GM EV1s, some who drove Toyota RAV4 electrics, and one couple who had an electric vehicle with solar panels on its roof. Those people were eager to tell what they liked and what they didn't like about their vehicles and to offer ideas about how an extended-range electric vehicle would be attractive to so many more people. Landy recalls that the couple with the solar roof saw an

Above: The bi-axial, in-wheel turbine braking system uses momentum and aero-thermal resistance to provide braking efficiency and to regenerate energy for the powertrain. Like the famed Chaparral 2J Can-Am racer, the 3E also has rear turbine extractors that provide downforce and battery cooling and also help with braking. *GM*

Left: Integrated thin-film polyvoltaic panels in the car's body turn sunlight into electricity. *GM*

Cristi Landy was the product planner for the Chevrolet Volt. Her job was to represent future customers as the car was being developed, to make sure the car had the equipment buyers of an electric vehicle with extended-range capability would expect on their vehicles. *GM*

EV with extended-range capability, such as the Volt, as a real game-changer that would spread the green gospel beyond what they termed as "the automotive vegans" who disdained the use of any imported petroleum. But what sort of features would people expect in a car like the Volt? Would they accept four seats instead of five? Would they be techno geeks who demanded 24/7 Internet connectivity? How simple would it have to be for them to plug the car in for recharging? Would they be willing to adjust their seats with manual controls, preferring that full electrical power be used to drive wheels instead of moving the driver's seat along its track? Those sorts of things were Landy's responsibility to bring to the table.

One of the things she fought for—and got—was the second display screen, the one in the center of the dashboard.

Instead of traditional gauges, the driver of the Volt has a screen display that shares information, such as speed, available charge in the battery pack, and fuel in the gasoline tank. But there was more information available than could be shown on that screen. Besides, that screen only was available to the driver's line of vision.

"You bring someone with you, the passengers in the car, and you want to show off. You need a 'brag screen,'" Landy contends.

And there was another aspect to what the second screen could do. "After you finish a trip, just like you do with a workout at the gym—you want to see that you burned three hundred calories and covered so many miles. Or in the case of the car, for this drive you covered forty-one miles, forty on electric power and one on gasoline, and you're at three hundred miles per gallon."

In keeping with Chevrolet's hallmark of affordability, Landy encouraged a solution

Above: Instead of the traditional cluster of gauges in front of the driver, the Chevrolet Volt presents information—from vehicle speed to range—on a computer-style screen. *GM*

Right: To monitor all of the Volt's systems and to present all the information a driver and passengers might want, the car has a second screen mounted at the top of the center stack. *GM*

Part of the space that might normally be taken up by a spare tire is filled with the cord and plug for recharging the Chevrolet Volt. *GM*

that would enable a recommendation of regular unleaded fuel for the generator engine. However, marketing lost that one because more expensive premium fuel allows the generator engine to run more efficiently, thus enhancing mileage.

Nonetheless, Landy knew passionate electric vehicle drivers saw an extended-range EV, such as the Volt, as a real game-change that would allow the EV movement to spread to the masses. While most consumers were not ready to accept the compromises and change in lifestyle required to drive a pure electric vehicle, the Volt's engineering design promoted flexibility through its extended-range capability, therefore making it much easier for more people to reduce their petroleum usage.

She also wanted a spare tire. Engineers objected on two fronts: They didn't want to add to the car's mass, and they needed the space usually occupied by the spare for some of the additional electronic components the Volt needed to carry.

There were similar debates over whether to offer a sunroof. The concept car had a clear roof, but such a feature might seem counterintuitive in a production vehicle because it could make the interior hotter on sunny days, thus demanding more from the air conditioning system and diverting power from the batteries that could be used to extend electric range.

At one point a "superfob" was considered, a key fob that resembled a small iPod with a display screen from which the Volt owner could manage various functions as well as unlocking the car's doors. Instead, General Motors' telemetrics/communication division, OnStar, worked up a series of special applications that can be accessed on a smart phone screen.

Through iPhone-style apps and a special Volt website, Volt owners will be able to monitor such things as battery power and recharging, receive turn-by-turn navigation for "green" routes designed to enhance electric mileage, receive—and share with other Volt owners—reports on their car's cumulative performance, and get tips after each trip about how they might have gotten even better mileage, etc.

OnStar also became part of a General Motors group working with utility companies and government regulators on so-called smart grid technology so Volts can be programmed—again, through OnStar apps—for recharging at the lowest possible electric rates.

OnStar also will be able to guide Volt owners to public charging stations as they travel.

In addition, OnStar worked with Volt development engineers to monitor several vehicle functions and to provide daily reports on vehicle diagnostics, not just on the Volt prototype a particular engineer was driving, but on all such cars being tested that day, thus providing an immediate overview that helped identify if an issue was isolated or something that needed wider attention.

In addition to developing uniform graphic displays for the Volt and its various apps, Stuart Norris' HMI design team was involved in the creation of

"You're making a significant change in your life buying this vehicle. This is not a usual car."
—GERY KISSEL

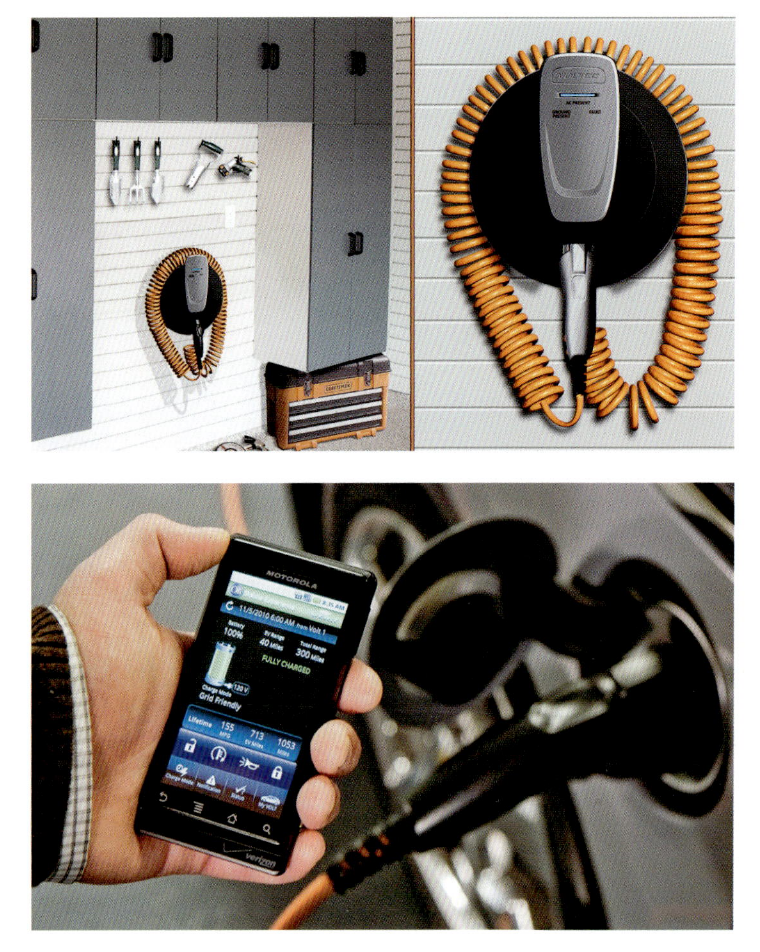

Top left: Garages equipped with 220-volt power get a special recharging unit for the Volt. *GM*

Bottom left: Smart phones can be used to manage and check on the Volt's state of recharging. *GM*

Above right: In addition to a filler for gasoline needed to run the generator engine, the Chevrolet Volt has a port for plugging in an electrical cord. *GM*

the hardware Gery Kissel and his team were developing for recharging the Volt's batteries.

"From a traditional vehicle design perspective, people focus on the exterior styling, the interior styling, the gesture of the vehicle, and so on," Norris says. But with a vehicle like the Volt, there was more to it because there would be more things with which the car owners would interact on a daily basis. Few people buy gasoline every day, but Volt owners are likely to plug in their vehicles each night.

"When we were doing the charging components, we did research," Norris says. "We went out. We photographed people's garages. What we really needed to find out is how customers use their garages. People's garages are there to keep all their stuff in. Where do they park their vehicles?

"You're making a significant change in your life buying this vehicle. This is not a usual car. There is so much at stake [that] you don't want to let the charging experience be a letdown.

"We've had to design the components so all the parts meet the requirements of sitting outside for ten years and not degrading. That's been a tough design challenge for us. You have to design around the fact that someone's probably going to leave the charge plug on the garage floor and drive over it."

Kissel not only worked on engineering the Volt's plug-in equipment, but he led the committee for the Society of Automotive Engineers who established the industry-wide standards for such equipment not only from a performance and safety standpoint but so that systems from various automakers—as well as standard and fast-charge equipment—will be compatible.

In SAE-speak, the standard is J1772.

With electric vehicles moving in the mainstream, Kissel recognized, there was a need for charging equipment that people would use daily, safely, and over a long period of time and would meet both SAE and UL (Underwriters Laboratory) standards.

Each Volt will come with a 20-foot cord that can be plugged into a 120-volt household outlet. There's also a 220-volt device that can be mounted to the garage wall for faster charging, or for charging, say, his and hers Volts.

"In either case," Kissel says, "there are lights that signal the recharge has begun."

"That's all you need to do," he says, adding that the system is so safe you can do that outside in the rain. "The standard we developed is suitable for outdoor application. It's sealed, and there are multiple safety redundancies built into the system. It's more than just a plug; it's an entire system."

Like so many others involved in Volt development, Kissel says that the EV1 gave GM a running start on what customer expectation is for charging. "We knew we had to keep it simple, and yet we also had to have features for advanced users or for specific time-of-day charging. But the bottom line was you had to have a mode to charge where it was simple. You plug it in and walk away."

The Volt is equipped with a 3.3-kilowatt charging system.

"People ask, 'If I get a Volt, what should I wire my garage to?' We give two answers," Kissel says. "If you want to wire for the Volt, it's a one-hundred-twenty-volt, fifteen-amp circuit breaker [pretty much the standard household outlet setup]. However, if you want to have the ability for the future vehicle coming down the road, or if you want two Volts in your garage, then you want two hundred forty volts at forty amps.

The 120-volt control box can be mounted on the wall and is designed for the Volt's cord to be wrapped around it when recharging isn't being done.

"The wall station looks like a wheel," Kissel says, adding that to his engineer's eye, "it looks prettier than it needs to be." But, he adds, "We wanted something that was styled and looked nice."

In some geographic locations, especially in warmer climates, many garages already have 240-volt wiring for clothes dryers or water heaters. And in some colder climates, some people with workshops in their garages also have such electrical systems.

Kissel says he expects charging will evolve over time, with hands-free charging systems in the future, with technology that allows you to pull up over a specially equipped parking place and simply walk away; the batteries in your car will recharge without needing any physical connection to an outlet.

There's also the emerging "smart grid," and, he adds, "probably something we haven't even thought of yet that will change the game, too."

"We want people to get in and say, 'This is different,'" product planner Landy says. "But we also want it to be easy to drive. We don't want people to have to say, 'Where's the shifter?' It should wow you for the technology but still be simple and intuitive."

And safe.

And in the case of the Volt, being safe involves not only protecting the vehicle's

One of Cristi Landy's jobs was to make sure the Volt fit into the Chevrolet dealership showroom. But Chevrolet isn't the only General Motors brand that figures to benefit from the Volt's technology. Here, Carl-Peter Forester (left) of GM Europe and Frank Weber (right) watch as Bob Lutz introduces the Opel Ampera, which will be sold in Europe with left-hand drive (by Opel) and with right-hand steering in England (by Vauxhall). *GM*

Preproduction versions of the Opel Ampera undergo testing. *GM*

occupants but also making sure that in the event of a crash, the high-voltage electrical equipment doesn't cause harm to occupants or first responders.

"Everything we do has another element compared to a traditional gasoline car," says Brian McGee, lead safety engineer for the Volt and based at General Motors' Milford Proving Grounds, where a number of Volt prototypes were crashed to validate the car's safety in various scenarios. McGee has mechanical engineering degrees from Oakland University and the University of Michigan and 30 years of experience at General Motors.

"We have one hundred years of experience in dealing with fuel systems—gasoline and diesel—and everyone is comfortable with them [despite their potential for fire after a crash]," says Larry Kwiecinski, Volt program "crash" manager and based at the GM Tech Center in Warren. Kwiecinski has mechanical engineering degrees from Wayne State and Rensselar Polytechnic Institute and more than 25 years at GM. "We want them to have the same comfort with the high-voltage electrical system. The experience may not be out there, but all the safeguards are being put into place."

For example, says McGee, in a serious crash, the battery pack is automatically isolated, disconnected by onboard computer systems. However, he added, "You don't want to disable the ability of a customer to get the car off the road in a fairly minor crash. You try not to shut down the high-voltage system in a drive-away condition. So we're trying to strike that balance, though the first consideration is protecting the customer."

Opposite: A concept for a Chevrolet Volt MPV5—a five-passenger crossover utility vehicle—was displayed at the 2010 Beijing International Automotive Exhibition. *GM*

Just as experience with the EV1 helped GM engineers as they developed the Volt, McGee says years of experience with GM gas/electric hybrids have helped provide insight into making sure the Volt's electrical system is neutralized in a crash.

Kwiecinski notes that McGee developed several new processes in preparation for crash-testing of the Volt.

"There are a lot of technicians and other people around these vehicles," he says. "We put a lot of processes in place to make sure our people were safe and that we'd get good data."

"We've learned a lot from hybrids," McGee says. "They plowed the ground. We've taken that and enlarged it. We're all electric, and the size of the components, the battery pack, and so on is much larger."

Kwiecinski and McGee and other safety engineers were involved in the Volt project from when the early designs were being laid out and the early specifications put together.

"We're staking out our ground to make sure we have protected structure for crashworthiness," adds Kwiecinski.

McGee says that while the Volt is built on GM's global compact car platform, "we don't think of it as a variant of the global small car. There are shared elements, but the Volt is so unique with the high-voltage battery and the place it needs to reside and the electrical drive unit, it's a very different car."

Long before there were prototypes for crashing, the safety engineers were doing computerized crash simulations on mathematical models of the Volt.

"There are things that only a physical test will uncover, things that a computer can't predict," McGee says. However, he adds, "We have not had an uh-ooh."

To some auto industry observers, it appeared the Volt suffered a series of uh-oohs when two key members of the development team—Robert Kruse, executive director for global vehicle engineering for hybrids, electric vehicles, and batteries; and Denise Gray, global director of recharageable energy storage systems—left General Motors. And then Frank Weber, who had overseen all Volt operations, returned to GM's European operations to oversee vehicle

Volt as luxury car? General Motors has seriously considered putting a Voltec powertrain into a vehicle based on the exotic Cadillac Converj concept. *GM*

development at Opel. There also was turnover at the top of Chevrolet brand management, and at the top of General Motors itself.

In the cases of Kruse and Gray, they had made their contributions to the Volt's development before they left. In Weber's case, he had taken the car from concept into prototype stage and was ready for a new challenge.

His return to Germany would not only help development of Opel's version of the Volt, the Ampera, but would allow Doug Parks to become the vehicle line executive of the Volt. Parks' specialty was taking vehicle programs into production, which was precisely the next step for the Volt.

"Doug will lead the total team through the productionization phase," Volt chief engineer Andrew Farah explains. "That's what he's good at. Frank was good at the early, creative stages."

"I'm a bit of a veteran and have launched a lot of programs," says Parks, GM's new global electric vehicle development executive. "I've been through the phase of validating them and bringing them into the plant seven or eight times."

Parks, who studied mechanical engineering at Michigan Tech University, joined General Motors in the mid-1980s and had done everything from interiors and systems engineering to ride and handling and noise and vibration control before moving into overall vehicle integration and then into being a vehicle development chief engineer, taking products, such as the Buick LeSabre, Oldsmobile Aurora, Pontiac Bonneville, Saturn Vue and Ion, Chevrolet Cobalt and HHR, and Pontiac Solstice and Saturn Sky, into production.

He was in Germany, heading engineering on GM's Delta platform and the development of the Opel Astra and Zafira and Chevrolet Cruze, when he returned to the United States in December 2009 to take the Volt into production.

"There's art and science to knowing when you're done," Parks says, adding that while a vehicle undergoes continuous improvement and updating throughout its lifespan, there's a point at which the engineers have to "lock in" and "finish up," turning their priorities from development to execution and quality assurance.

"I was excited that General Motors decided to step out in front and say we're going to lead," Parks says of the decision to showcase an EV with extended-range capability, and not just to showcase the idea as a concept but to actually take the car into production and to do it the right way, the new way he sees GM doing its vehicle development.

"In the past, on a lot of programs I worked on, we had cost targets to stay within and they were balanced for the business case, and for good reasons. But when we tore our competitors' vehicles apart to do benchmarking, they had spent more money on the hardware and on material quality that leads to tactile feel that customers perceive as real value. It was frustrating.

"But starting with the Volt, and even with the global Delta vehicles, somewhere around 2005 or 2006, we started realizing what was more important was to bring out a car that was going to beat the competition. We had to do that first and to make the business case work second."

But, he adds, now that the company has emerged from bankruptcy, "we can have an acceptable business case and put a lot more designed and engineered content into the car, and the business case still works.

"Before, it was 'how good of a vehicle can you do to meet this business case?' Now, it's the other way around. The Volt may not make money right now, but it's a great vehicle and it's a stepping-out-in-front-of-technology vehicle, a world-leading vehicle, and we'll figure out how to make the business case work. We're taking the gloves off and doing the best that GM knows how to do."

Andrew Farah notes that despite General Motors' financial woes—the company went through bankruptcy while the Volt was being developed—the program never lacked for resources. "I have never been told: 'Thou shalt cut X,' but rather I was called into a room and asked . . . 'What do you need?'" Farah says.

Parks says he sees that same sea change throughout GM's product development effort.

"The programs that made the cut [survived bankruptcy], the company's investing in them significantly," he says, adding, for example, that "the Chevy Cruze is the best [small] car we know how to do. We didn't hold back."

Neither would those working on the Volt.

Shaken, not stirred. A special stand shakes a Chevrolet Volt prototype through a year's worth of wear and tear as part of the car's development and preproduction validation. *GM*

An x-ray view shows through the sheetmetal skin of the Chevrolet Volt. *GM*

CHAPTER 9

WITH VOLT, DETROIT BECOMES THE (ELECTRIC) MOTOR CITY

WHEN PRODUCTION OF THE 2011 CHEVROLET VOLT BEGAN, THE LITHIUM-ION BATTERIES THAT PROVIDE IT WITH SOME 40 MILES OF GASOLINE-FREE RANGE WERE PRODUCED IN SOUTH KOREA. BUT THAT WAS SCHEDULED TO CHANGE—AND SOON—AS CONSTRUCTION BEGAN ON LG CHEM'S NEW $303-MILLION BATTERY PRODUCTION FACILITY IN HOLLAND, A CITY IN WESTERN MICHIGAN KNOWN FOR ITS DUTCH HERITAGE, ITS HAND-SCRUBBED STREETS, AND A RAINBOW-COLORFUL TULIP TIME FESTIVAL THAT EACH SPRING ATTRACTS MORE THAN HALF-A-MILLION VISITORS.

The first preproduction or "pilot" Chevrolet Volt begins moving along the assembly line in General Motors' Detroit-Hamtramck Assembly Plant. *GM*

When that new lithium-ion battery cell plant comes online in 2012 and starts shipping its cells not across the ocean but across the state to the GM Brownstown Battery Assembly Plant, the Chevrolet Volt not only will be the first EV with extended-range capability in the world but perhaps the most "made-in-Michigan." Its battery packs, engine generator, sheetmetal production, and final vehicle assembly—even the source of its "fuel"—all will be produced in the mitten-shaped state that put the world on wheels a century ago.

With the Chevrolet Volt, Detroit again reclaims its title as the Motor City, though in this case it's the Electric Motor City.

Final assembly of the Volt takes place within Detroit, though not only in the actual city limits. DHAM, as it's known within General Motors, or Detroit-Hamtramck Assembly Plant, as it says on the sign outside the huge facility, is located in Hamtramck, a city surrounded on three sides by the city of Detroit. Its western border is another "island" city, Highland Park, where Henry Ford built his Model Ts. Hamtramck was organized as a township in the eighteenth century, incorporated as a village in 1901, and became its own city in 1922, eight years after the Dodge brothers opened their famed Dodge Main assembly plant there.

The Dodge brothers opened their plant in 1910 to build engines for Henry Ford and other early Detroit automakers. They expanded the facility in 1914 when they began producing cars under their own name. The plant remained in production until 1980.

In 1985, as part of a massive urban renewal project, General Motors opened its 3.6-million-square-foot DHAM Assembly Plant on a site with a footprint so large that it extends across both the Detroit and Hamtramck city boundaries, includes its own power plant, 24 miles of a conveyor system, a 16.5-acre wildlife habitat conservation area, and Hamtramck's only cemetery. At any given time, as many as 1,600 vehicles can be somewhere in process at DHAM.

Originally, DHAM built the Buick Riviera, Oldsmobile Toronado, and Cadillac Eldorado and Seville. In 1987 it stretched its assembly line across the Atlantic Ocean to build the Cadillac Allante. In 2010, the Chevrolet Volt began moving along a line on which the United Auto Workers' Local 22 also assembled the Buick Lucerne and Cadillac DTS luxury sedans.

"That's what we have expertise in—in building cars," says Theresa "Teri" Quigley, DHAM's plant manager, who notes that the average UAW member in her facility has a quarter-century of car-building experience.

"I want to capitalize on those many, many, many years of building cars instead of making this [car] seem like something totally different," she adds.

However, she says, DHAM also has some newer employees. So to get everyone to the same baseline, and to prepare for the visibility the first extended-range electric vehicle brings to the plant, management decided to reconstruct the plant to make sure all 1,100 employees would be familiar and comfortable in dealing with their new product's new technology, especially its high-voltage technology, and in the latest in General Motors car-building processes that are being installed in plants around the world.

"A lot of people were intimidated by the sound of it—Volt."
—LARRY JONES

"A lot of people were intimidated by the sound of it—Volt," says Larry Jones, a UAW Local 22 member who spent a lot of time working with those who built the first Volt prototypes—the IV cars—in the Pre-Production Operations (PPO) shop at the GM Tech Center in Warren. Jones and others from DHAM not only learned about the Volt early on, but they were able to help those in PPO learn about how cars are built not one at a time but on a running assembly line.

Instead of the flow of an assembly plant, Jones says, the goal of the IV builds is to get the cars done as quickly as possible so engineers can begin testing, even if it means some parts have to be removed so others can be installed later in the process.

"We showed them how you could flow the build process," he says.

DHAM also worked with Volt design engineers on such things as color-coding some parts to avoid possible confusion during the assembly process.

The first battery pack produced at General Motors' Brownstown Township Battery Assembly Plant makes its way through the facility. *GM*

Jones and others from DHAM prepared documentation about the Volt assembly process that would be needed back at their plant. They also helped set up a work station in which every Local 22 operator could become familiar with the car long before the first Volt started moving along the assembly line.

Thus, in an alcove deep within the plant, a Volt prototype is built up over and over again with each operator putting hands on the car, installing the same component(s) they'll do on the line, but with time to see how each piece fits and integrates, to learn when the battery pack is "live" and when it is de-energized, and to deal with any fears he or she may have had about the car and its electrical systems long before those cars are actually moving down the line.

"Now they understand it," Larry Jones says. "It feels like any other car. They're not intimidated."

"We've been proactive," Teri Quigley adds. "We've said, 'We're going to arm you up front with all the information, the safety knowledge,' and have said that you need to raise your hand if you're not comfortable yet and we'll go back through all the training before we ever get cars on the line."

Once upon a time, Quigley says, the attitude in the plant was, "I'll do as I am told and no more." Now, however, she says, team leaders and team members feel real ownership and take real pride in their work.

"We are proud of what we do here," she says. "At the end of the day, I'm a landlord here, and the safety of these folks is my responsibility."

Gratefully, she adds, "I am not sensing apprehension, and that's because of all the training we've done. And I expect that we'll do some reiteration because we're going to learn more as we go."

Quigley has been at DHAM since 2006, when she arrived as assistant plant manager. She became plant manager a year later. The only daughter in a large farm family, she was always mechanically inclined and was the first girl in her high school ever to enroll in the vocational auto shop class. When she graduated, she not only had a high school diploma but a certification in engine and transmission repair. She wasn't sure about her career goals until General Motors Institute (now Kettering University) began offering an engineering degree in manufacturing systems. She enrolled and later went on to earn a pair of master's degrees as well.

"I was always the one in engineering school who was way more right-brained," she says. "We'd do those kite diagrams that show where your likes and dislikes were, and we'd have to line up by our number, and I'd be outside in the hallway.

"I really like interacting with people—and you have a lot of people to interact with when you're building a car." Walk around the plant with Quigley, and you immediately notice that she seems to know all 1,100 employees on a first-name basis, and she often stops to ask about family members as well.

"I don't think I could survive well in the engineering world if I was just doing engineering. I would starve from a lack of people interaction," she says.

Quigley started working as a co-op student at GM's Pontiac West plant. She spent a decade at the company's assembly plant in Moraine, Ohio, working in seemingly every plant management role—from the paint shop to general assembly to assistant plant manager. She then went overseas—to a GM plant in England for two years. She returned as assistant plant manager at Spring Hill, Tennessee, and then moved back to Michigan to work at DHAM.

For Quigley and DHAM, work on the Volt began more than a year before production began. And a year before that, GM production engineers started doing computerized virtual Volt assembly, with demolition and the start of reconstruction and new equipment installation timed to coincide with the annual Christmas holiday vacation period. Quigley notes that cooperation not only from UAW Local 22, but also from other UAW locals at other GM facilities allowed some equipment to be moved between GM facilities, saving time and money, perhaps as much as 11 months and $40 million.

Sheetmetal components for the Volt are not only produced at DHAM but also come from GM stamping plants in Lansing and Pontiac. Bodies are produced within the DHAM body shop and then go to the paint shop, where things get complicated because of the Volt's concept-style black roof, which requires the body to be painted and then masked off so the roof can be resprayed in black.

The Volt bodies are placed on the same final assembly line as the Lucerne and DTS, with cars intermingled to meet their respective production numbers and to keep the line moving at a steady rate.

Battery plant operations manager Marisol McCormick delivers the first Chevrolet Volt battery pack off the assembly line on January 7, 2010. *GM*

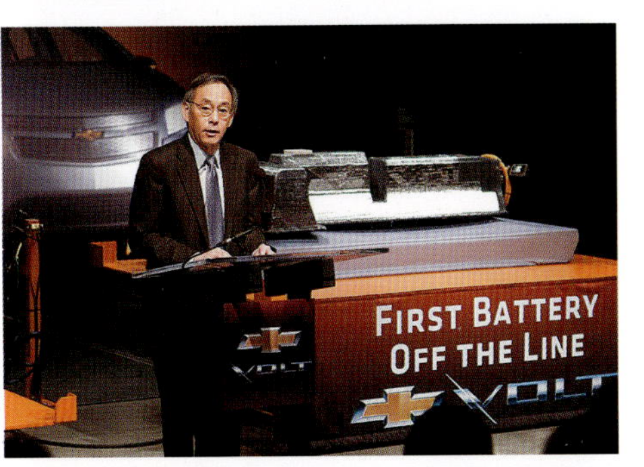

U.S. Energy Secretary Steven Chu speaks at an event held to celebrate the delivery of the first battery pack from the Brownstown facility. *GM*

General Motors chairman Ed Whitacre congratulates Brownstown plant manager Nancy Laubenthal on the delivery of the plant's first Chevrolet Volt battery pack. *GM*

The first preproduction Volt moves from the body shop into the assembly area at Detroit-Hamtramck (DHAM). The first build took place in late March 2010, as the plant geared up for full production in November. *GM*

The Volt prototype takes its place on the assembly line behind other vehicles being built at DHAM, where the Cadillac DTS and Buick Lucerne also are being assembled. *GM*

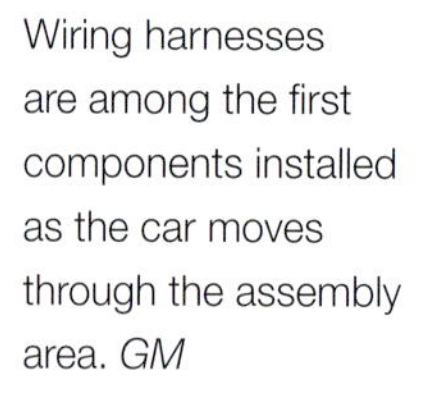

Wiring harnesses are among the first components installed as the car moves through the assembly area. *GM*

Work areas are well lit to help United Auto Workers (UAW) members properly fit components into and onto the vehicle. *GM*

The Chevrolet Volt presents assembly workers with even more hoses and electrical lines than the typical automobile. *GM*

Chevrolet Volt chief engineer Andrew Farah and DHAM plant manager Teri Quigley look pleased with the way things are going as the first Volt makes its way along the assembly line. *GM*

With a new vehicle on the line, sometimes engineers outnumber UAW workers. *GM*

Eventually, the Lucerne and DTS will be "built out," and other GM electric vehicles will join the Volt on the DHAM line.

The Volt's assembly is unique at several stations along the line. For example, the Volt has four cooling systems and the affiliated hoses and electric controls—for occupants, for the battery pack, and for the engine/generator and power electronics. The battery pack itself is a large T-shaped structure that bolts to the bottom of the car.

Quigley says that while the Volt has an electric powerplant, its assembly isn't radically different than building a car with a traditional internal combustion power system.

"Without such components," she says, "we're just building expensive furniture."

The battery packs for the Volt are produced less than 35 miles from DHAM, just down Interstate 75 in an industrial park in the suburb of Brownstown Township, on what used to be the site of Detroit Dragway.

"The assembly process involves three lines," says Nancy Laubenthal, plant manager of the Brownstown Township battery pack assembly plant.

The first line, which is highly automated, starts with the lithium-ion cells. They are about the size of a typical automotive license plate, she says. "This is a pre-assembly area where we basically take the lithium-ion cells and other components and bring them together in one of three modules," Laubenthal explains. "The battery pack has a T shape. One module goes into the cross of the T, and the long part has two modules."

The second line at Brownstown Township is module final assembly, which involves a lot of welding—"Three-hundred-some welds on every battery pack," Laubenthal says—as well as module verification, leak testing, and electrical charging and mechanical completion.

"The last line is called the pack main line," Laubenthal says. "We take those three modules and assemble them into a battery pack with mechanical and electrical connections. There's a lot of testing of the battery pack on that line as well." That testing, she says, involves dimensional validation, leak performance, and thermal response.

Each pack is about five feet long and weighs some 400 pounds.

Laubenthal joined General Motors as a co-op student in the 1980s. She has multiple degrees in chemistry and has spent much of her career at GM working with paint and polymers.

She says one major difference between other aspects of GM assembly and the establishment of the battery pack production facility was building something totally new.

"When I worked at Lansing–Grand River and we were building a new plant, we went to visit other assembly plants that were particularly successful, to learn the best practices," she says. But Brownstown Township is the first facility of its kind. "We're creating the best practices," she says. "That's exciting and challenging at the same time. We're creating the cycles and capturing the cycles of learning for this plant and for future plants in battery assembly."

Laubenthal adds that the third line reminds her of a general assembly line in an auto assembly plant. Not only are the packs moved by automated vehicle at that point, but there are connections to be made, electrical wiring to be done, hoses to be installed, and verification testing to do. "At the end of the pack main line, the pack is completely assembled, covered, and ready to ship to Detroit-Hamtramck," she says.

At DHAM, the battery pack is attached to the underside of the car body as it moves along the line.

The highlight of any car's assembly is the so-called marriage ceremony, where the powertrain is inserted from beneath the chassis.

"The marriage will look the same," Quigley says.

But while it might look the same, the way the wheels of the Chevrolet Volt receive their power will be very different. That power will come not from purely mechanical movement but from electrical connections.

Even while going around a corner in the plant, the Volt comes under close examination. *GM*

Hans Kaiser climbs inside a Chevrolet Volt to make to verify installation of components as the car moves along the assembly line. *GM*

Above: The Volt rides through the assembly process on an automated cart, which is designed to carry along many of the components to be placed on the car. The process makes assembly easier and helps assure each car gets the specific parts assigned to it. *GM*

Opposite top: (From left), Adrian James, Fred Donaldson, and Lezie Stewert, UAW and salaried staff at the DHAM assembly plant, make sure everything is progressing properly as a Chevrolet Volt works its way down the line. *GM*

Opposite bottom: Kathy Adams (left) and Carry West install rubber window trim on the Chevrolet Volt. *GM*

Above: Automated lifts help with the installation of large components, such as the dashboard assembly. *GM*

Left: With its instrument panel in place, the Volt moves along the line. *GM*

Above: Every part of every car has to be designed, engineered, validated, produced, and assembled, and then installed on the vehicle. *GM*

RATED LOAD 330 LBS.

Above: The addition of any new vehicle to an assembly line draws a crowd, and even more so when that vehicle is as revolutionary as the Chevrolet Volt. *GM*

Left: The plug-in port where the Volt's batteries will be recharged. *GM*

Opposite: Accordion-style lifts help workers marry the powertrain components with the chassis. *GM*

The battery pack is prepped for its installation. *GM*

The battery pack is carefully moved into position and secured to the bottom of the chassis. *GM*

It's also a good sign when the lights work. *GM*

Chief engineer Andrew Farah gets to plug in the first Volt to energize the battery pack. *GM*

Everything is inspected and then re-inspected. *GM*

Paper now stuck to the front bumper shows details of the components scheduled for installation. In the lower right-hand corner, the words read "Quality Creates Customer Enthusiasm." *GM*

Wheels and tires are installed. *GM*

Andrew Farah rides along as Teri Quigley gets ready to drive the first Volt off the assembly line. *GM*

DHAM Assembly Plant employees celebrate as the first Volt is driven off the line. *GM*

Although the preproduction cars won't be sold through dealerships, they'll be used for extensive testing and will rack up hundreds of thousands of miles during validation drives by GM staffers. *GM*

CHAPTER 10

Driven to Be So Extraordinary That It's Downright and Delightfully Ordinary

On May 22, 2008, Bob Lutz climbed out from behind the steering wheel of a previous-generation Chevrolet Malibu with a grin on his face that was wider than any Malibu owner might understand. Lutz, the vice chairman of General Motors, walked around to the car's hood and used a permanent marker pen to sign his autograph and to add this statement:

"We are making history today."

That was Lutz' reaction to his first drive in one of the first running prototypes propelled by a preliminary version of the powertrain that would be used in the 2011 Chevrolet Volt.

Thirty months earlier, Lutz, the company's most senior and well-known "car guy," had suggested General Motors do an electric-powered concept car, and not just any electric-powered car but one equipped with lithium-ion batteries, the sort normally used for laptop computers and cellular phones.

Lutz' idea was far from popular within the cylindrical towers of General Motors' headquarters in downtown Detroit or at the campuslike GM Technical Center in north suburban Warren, Michigan. In fact, Jon Lauckner, GM vice president for global (vehicle development) program management, told Lutz that while his idea may not have been dumb, it wasn't smart either.

However, unlike so many others, Lauckner liked what Lutz was wanting to do; he just wanted to modify Lutz' idea, to do a concept that would, indeed, run on electricity but would carry a small internal-combustion engine as an onboard generator to refresh energy to the battery pack and thus greatly extend the vehicle's viable driving range from tens to hundreds of miles.

Lutz' idea, Lauckner's technical twist, and a lion-inspired design combined to produce the Chevrolet Volt concept, unveiled at the 2007 North American International Auto Show just a couple of blocks down Jefferson Avenue from GM headquarters. Not long after that show, where the Volt was the big news of the first major international automotive event of the year, the General Motors board of directors encouraged engineering development and needed battery science to take place as preparation for the concept's potential production.

To prove that those preparations were progressing properly, in the spring of 2008 three old Chevrolet Malibu body shells were outfitted with prototype extended-range electric powertrains for preliminary testing at General Motors' secure proving grounds just west of the quaint Midwestern village of Milford, Michigan. That's where Lutz drove and showed his obvious pleasure with the progress.

As that summer progressed, 20 more prototypes were built, this time using Korean-produced J300 bodies that were based on a Daewoo sedan built on the General Motors global compact car architecture that would be used for the anticipated Volt production car.

By the summer of 2009, the first of some 80 true Volt prototypes, cars based on the intended structure and componentry designed specifically for the Volt, were working their way down an assembly line in the Pre-Production Operations (PPO) center within the GM Technical Center campus. These cars, known within General Motors as IVERs (integration vehicle engineering release), would be used to test production-intent equipment and systems before being destroyed during crash testing.

Opposite and below: On May 22, 2008, Bob Lutz did his first drive in a car equipped with the extended-range electric powertrain proposed for the Chevrolet Volt. Obviously, he was pleased with the experience. *GM*

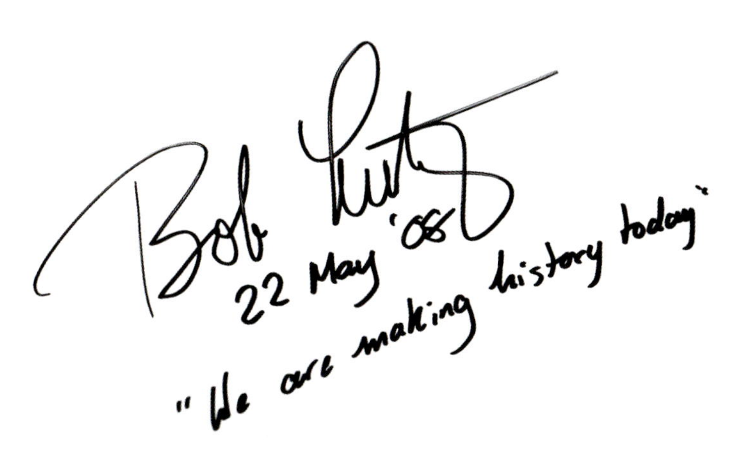

Bob Lutz
22 May '08
"We are making history today"

The extended-range electric powertrain for the Chevrolet Volt underwent dynamic development in what are termed "mules" in the auto industry. For the Volt, those cars were either previous-generation Chevrolet Malibus or cars produced by General Motors' Korean arm that shared basic architecture with the Volt chassis that was being designed. Here, one of the mules drives past the water tower and lake at the GM Tech Center in Warren, Michigan. *GM*

Volt mules were used not only for internal testing by GM, but to let news media, government officials, and others in the electric vehicle industry get a feel for the Volt's extended-range electric powertrain. *GM*

Then, in the spring of 2010, actual preproduction versions of the Volt began to trickle out of the Detroit-Hamtramck Assembly Plant as part of the ramp up to the start of Volt production in November.

But even before there are any cars to be driven, General Motors vehicle development programs drive toward production through a series of "gates," crucial go/no-go stages, such as DSI, the document of strategic intent; VPI, vehicle program initiative; DSO, design sign-off; and VDR, verified data release. Pass successfully through VDR and early prototypes (mules, as they're known in the industry), and IVERs start to roll, first toward the test track and at some point out onto public roads—though all the time wearing camouflage of various sorts to hide all the details of the production-intent design.

For General Motors vehicle programs, the first big test on public roads for IVER prototypes is known as the 65-Percent Drive, the 65-percent figure indicating that vehicles are well beyond halfway toward production readiness. The 65-Percent Drive involves several cars, and competitive vehicles (if such vehicles exist), and twice as many engineers as cars so they can poke and prod, critique and dissect how the vehicles and their various systems are coming along—what's good, what can be even better, and what needs priority with regard to improvement.

For the Chevrolet Volt, the 65-Percent Drive started at the Milford Proving Grounds northwest of Detroit and followed a route south through Toledo and then eastward across Ohio to Pittsburgh (and then back).

Opposite: Can a mule pose for a beauty shot? It can when the trees are in bloom at the GM Tech Center campus. *GM*

VOLT
TECHNOLOGY INSIDE

A Volt mule takes a test drive behind GM's downtown Detroit headquarters, where OnStar has its research laboratory. *GM*

The Chevrolet Volt engineering team gets ready to leave General Motors' Milford Proving Grounds on the 65-Percent Drive, the first major road trip for preproduction cars. *GM*

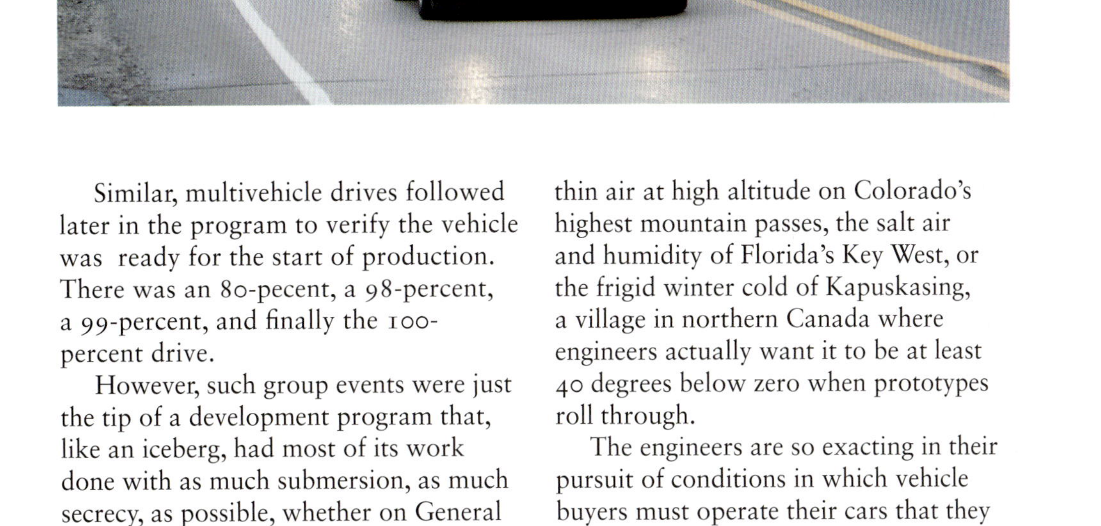

On the road, and with only minimal camouflage. *GM*

Similar, multivehicle drives followed later in the program to verify the vehicle was ready for the start of production. There was an 80-pecent, a 98-percent, a 99-percent, and finally the 100-percent drive.

However, such group events were just the tip of a development program that, like an iceberg, had most of its work done with as much submersion, as much secrecy, as possible, whether on General Motors' test tracks or by sending a disguised car or two to experience extreme weather and climate conditions. Some of those conditions included trips through the scorching heat and steep hills that challenge you as you drive across and then climb out of Death Valley in the middle of summer, or the cold, thin air at high altitude on Colorado's highest mountain passes, the salt air and humidity of Florida's Key West, or the frigid winter cold of Kapuskasing, a village in northern Canada where engineers actually want it to be at least 40 degrees below zero when prototypes roll through.

The engineers are so exacting in their pursuit of conditions in which vehicle buyers must operate their cars that they even seek out areas, such as those in the Midwestern United States, where there are magnetic anomalies and convergences that might interfere with or confuse a vehicle's onboard computerized systems.

Overcoming such challenges are the sort of obstacles automotive development engineers face on a daily basis.

Top: Engineers use computers to verify that all the car's electronics are functioning as designed. *GM*

Bottom: From Milford, the cars drove southeast through Michigan and then into Ohio. *GM*

"This car has had every challenge that every car has, and then some," Volt chief engineer Andrew Farah said as development work continued in the weeks leading up to the start of production.

For example, Farah says, altitude matters when it comes to the Volt, though not necessarily in the same way it does for a conventional car.

Conventional cars suffer significant power losses as they climb to high altitudes, where the air is thin and combustion becomes starved for oxygen. That's not the case when the Volt is operating under battery power. However, Farah says, "Hills are an issue, because they can deplete the battery more quickly as you climb. And when you enter extended range at altitude, the gasoline-powered generator doesn't put out as much power as it does at normal elevations. So what happens is the battery has to participate more."

Complicating the challenge is the fact that in mountainous terrain, the road is rarely flat.

"You're usually going up and down. It's not like you get up in the mountains and it's flat," Farah says. "You're going up or down. That being the case, the engine participates more often, and the battery needs more participation."

An additional complication: The driver and passenger want to be comfortable as they travel those mountain roads, where they can encounter extremes in weather conditions, including summertime snowstorms. So they use the heater or the air conditioning, and that can put even more demand on the battery.

"So mountains are a problem," Farah says. But while they were a literal large obstacle to overcome for the Volt development team, they were, in fact, just one more item to account for on the engineering group's to-do list.

"I was recently part of a team that took several Chevrolet Volt preproduction vehicles to the Pikes Peak Summit House [altitude: 14,110 feet above sea level] in Colorado," John Blanchard, lead calibration engineer for the Volt engine-generator, wrote in a blog on GM's Voltage website more than a year before the start of production.

"Getting to the Summit House involved driving the Volt on a partially paved, twisting nineteen-mile highway. It's a great test for any vehicle. Our team used this trip to evaluate a number of aspects of the Volt. On the trip up, we were making sure the Volt could climb the steep inclines and operate at a high altitude. The Volt was in extended-range mode for the most part of this segment. On the downhill segment, we were examining how well the regenerating [braking] feature of the Volt was adding electrical energy back into the lithium-ion battery.

"We were pleased with the results on both segments. The Volt climbed the mountain faster than we anticipated, and the regenerative feature produced a good amount of energy back into the battery. We were also pleasantly surprised with the temperature of our brakes. The National Park Service [which checks vehicles' brake temperatures as they descend] at Pikes Peak said it was one of the coolest temperatures on brakes that they had ever seen."

That drive occurred just a few weeks after engineers took the Volt up what they consider the most challenging driving hills east of the Mississippi. "Driving the twisty, winding roads of Knoxville, Tennessee, you really get to see what a car is made of," blogged Alex Cattelan, powertrain assistant chief engineer for the Voltec electric propulsion system. "As we all know, it's not enough that a car have the technical engineering to operate seamlessly, but it has to have the right feel, too. The vehicle reaction needs to be intuitive to the driver. Whether driving in pure electric mode with a charged battery or when the engine kicks in to sustain battery charge, the car must always have the same responsiveness that a driver would expect from any great vehicle. Push on the throttle, the car speeds up. Drive a steep grade, the car makes the climb. The thermometer outside reads ninety-five degrees, the vehicle takes the heat.

"It's my team's responsibility to ensure that the drive performance of the next-generation electric vehicle is meeting expectations. To do that, our team drove seven test vehicles from Milford, Michigan, to the mountains of Tennessee to calibrate how the Volt's battery energy, fuel efficiency, and drive quality work together in real-world conditions. Right now, our focus is on the driver experience, and I have to say we are happy with the results. After testing the Volt in cold, winter conditions in Canada, in high altitudes in Denver, in hot climates in Death Valley, and now in Knoxville, we have not found any surprises. "A baseline has been established that we can use to fine tune further development. Cabin conditions and under-hood temperatures all stand up to the heat and grade challenges put to the battery pack. System testing to date verifies that we can properly balance vehicle requirements, such as drive performance, drive feel, thermal conditions, and efficiency. Everything we are doing proves the Volt is right on track."

By the way, she adds, "I look forward to hitting the open road again—but this time on my Suzuki GSXR 750 motorcycle."

Volt prototypes cross the spectacular Veterans' Glass City Skyway bridge in Toledo. *GM*

Pikes Peak and the Tennessee hills, two real-world driving situations Volt prototypes were tested in. Yet engineers also had to prepare the cars for situations no real-world driver might ever try, Farah notes. As he puts it, they have to make sure the car is ready for "the lunatic fringe." Thus Farah offers what he calls "the infinite six-percent grade":

"What happens is you leave your house fully charged and drive up a six-percent infinite grade, and it just keeps going, forever and ever and ever. What happens?" he asks.

"Let's say you pick a speed. I'm going to pick seventy miles per hour. You're probably going to require more than sixty kilowatts to do this.

"I have no problem getting sixty kilowatts out of that battery, and I get it for some period of time, but not for forty miles. But nonetheless, I'm doing just fine.

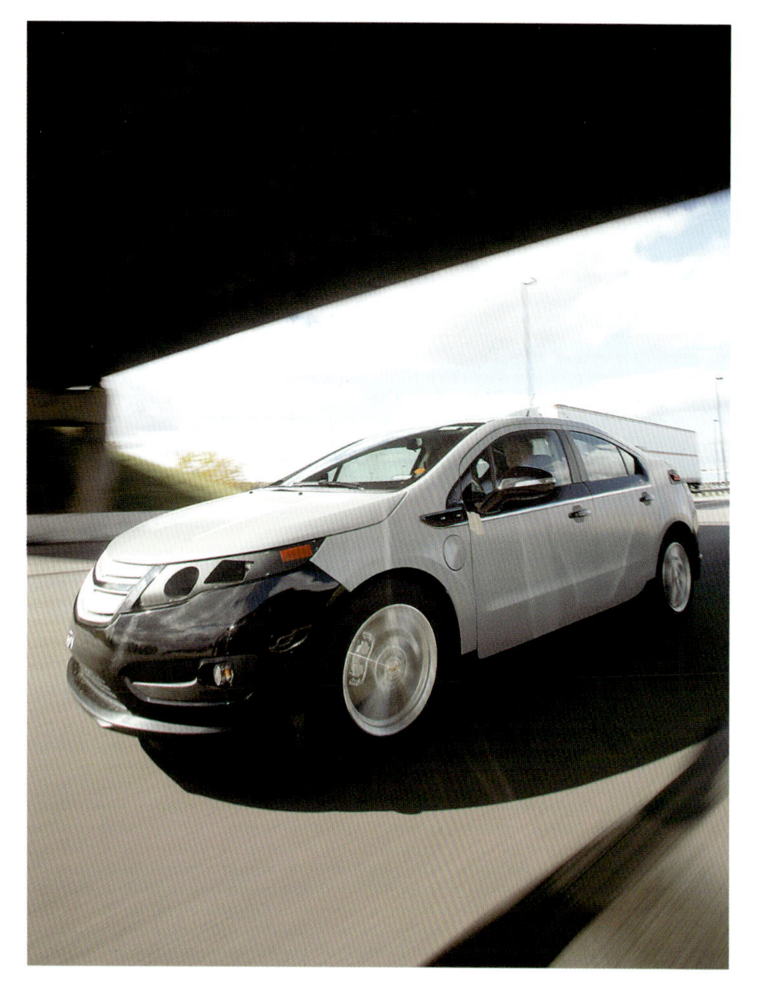

"At some point the engine kicks in, but the engine can only deliver, let's say, fifty kilowatts, and that figure is falling because of the altitude. So what will happen is at some point, the vehicle speed will come down, and you will restabilize at what the engine can make up because the battery will not be able to participate anymore.

"So, what's your terminal velocity?"

As it turns out, even in such lunatic fringe scenarios, the vehicle keeps going at more than 55 miles per hour for many miles, so you can climb the mountains with the car.

Farah remembers going on a ski trip in the '80s with three buddies in an imported Pontiac LeMans—four guys with a ski rack on the car and having to keep the throttle wide open to climb up to the ski hills. "We were lucky if we could do 50."

"People who live in the mountains know this. They understand it, and they deal with it," he says.

To make it easier for those who drive in mountainous areas to deal with their elevated challenge—and to help those who live in a very different sort of environment deal with yet another unusual driving situation—the Volt

engineering team created "mountain" and "hold" driving modes and integrated their activation into a switch on the center console that also allows the driver to select between normal and sport modes.

Many Volt owners will leave the control in its normal mode for all of their driving. Some will sometimes switch to sport mode for even more athletic, aggressive driving dynamics.

In mountain mode, the Volt's energy control computer kicks on the onboard generator engine a bit sooner to make sure the battery pack has enough charge at all times so the car can cruise, for example, up to Colorado's West Loveland Pass and through the Eisenhower Tunnel on Interstate 70 at the posted speed limits, as opposed to the driver having to worry about Farah's infinite grade gymnastics. At an elevation of more than 11,000 feet, the tunnel is the highest in the world designed for vehicular traffic.

"Going up and down the mountains, in climates and altitude, those are environments that a lot of alternative propulsion vehicles wouldn't want to tackle, but we're making sure someone who lives in a ski area of Colorado could buy a Volt and be perfectly satisfied," says Pam Fletcher, General Motors global Voltec and plug-in hybrid electric powertrain chief engineer.

"We have new modes of operation, and we've figured out how to be the most efficient and still be pleasing to the customer," she adds, noting the challenge of integrating electric motors and the battery pack and the onboard gasoline-powered generator so the transitions remain seamless in all driving situations.

"It's uneventful, and that's what we strive for," says Fletcher, who got into engineering in part because her father raced cars, and since she had no brothers, she helped her father work on his cars. She studied mechanical

The first, big on-public-roads excursion for General Motors prototype vehicles is the so-called 65-percent drive, which takes a group of production-intent prototype vehicles on a long drive. For the Volt, that drive started at the GM Proving Grounds in Milford, Michigan, went south into Ohio and then east at Pittsburgh and back. On the drive, the cars are exposed to real-world conditions and traffic, with engineers monitoring every aspect of the vehicle performance. *GM*

The leaves have turned to their autumn colors as the Volts drive into Pennsylvania. *GM*

engineering at the General Motors Institute and spent summers working in a manufacturing plant, which she quickly realized was not what she wanted to do for the rest of her life.

Instead, she landed a job at specialty engine-builder McLaren, which built high-performance engines for cars like the Buick GNX.

"It was during the advent of electronic controls for engines and transmissions, and that intrigued me," she says.

After a short stint doing engine calibration work at Ford, Fletcher returned to General Motors, became chief engineer for the Chevrolet TrailBlazer and GMC Envoy, but then left GM again when her husband was offered a job in North Carolina.

Armed with a master's degree in thermal sciences, she eventually returned to GM to work on hybrids and now electric vehicles.

"After one hundred twenty years of internal combustion propulsion, things are changing," she says.

But even as things change, they still have to obey what Volt chief engineer Andrew Farah refers to as "the Larry Laws of Physics."

Larry Laws has been a General Motors engineer since 1991. For the Volt, he's the EDQE—energy and drive quality engineer—which means he uses physics to model "pretty much how the car behaves in various conditions," whether urban centers or mountains or parts in between.

Laws and his laws are crucial to discussions about what tradeoffs can and should be made as engineers work to optimize vehicle efficiency. He also becomes involved in validating the results on the road.

Farah simply refers to his role as "the long arm of the Laws."

"The effort is validating the model and doing the diligence and the controls," Laws says. "Where we thought we would be and where we are is very close—the model and the test data are within about five percent of each other, and there's an upside we didn't know we had."

Laws was energy engineer for the EV1 and worked on energy modeling and efficiency on fuel cell, hybrid, and other alternative-fuel powertrains before joining the Volt team.

He jokes that the three unbreakable laws of thermodynamics are the following:

1. *You cannot win.*
2. *You cannot tie.*
3. *You have to play.*

While the real laws of thermodynamics cannot be broken, Laws says that with vehicles with hybrid or Volt-style extended-range powertrains, "you can move the energy around more efficiently, and that's the pseudo rule-breaking we're trying to do, to use less energy to get more output."

With the Volt, he says, "we can make tweaks or optimizations to make the first forty miles as good as possible, as efficient as possible."

For example, he says, you could close off the front of the car to enhance aerodynamic numbers and extend electric range, but you also need to supply airflow for powertrain cooling when the generator is running. The challenge is to find the right balance for efficiency and durability.

Note that the engineers don't use the word *compromise*.

"I think it's a good blend," Farah says, discussing vehicle systems development. "I'm not going to call it a compromise because that assumes somebody didn't do something that was really important. We did, but we tempered it appropriately."

"We're always working to make it better and better," Laws explains. "We're always trying to take every unnecessary use of watts out of the car, and we're always discovering new things."

For example, the Volt has a battery-only range of around 40 miles at typical city-driving speeds. But how far can you get if you put your foot to the floor and run at 100 miles per hour on an open road? How soon do you exhaust the batteries and force the generator to start resupplying energy to the battery pack?

"Aerodynamic losses are cubed with velocity," Laws reveals, noting that while at 50 miles per hour you're drawing 10 kilowatts of energy from the batteries, at 100 miles per hour, energy drain soars to 40 kilowatts. Nonetheless, he adds, even with your foot floored to 100 miles per hour, you still should be able to travel around 22 miles before the engine generator is needed. Simply drive aggressively, rather than at such a reckless speed, and keep the air conditioning off, and you should be able to cover 30 miles on batteries only, he adds.

Farah adds that if you achieve those 20-plus miles at 100 miles per hour, you've done so while using the energy equivalent of only half a gallon of gasoline.

Such are the equations development engineers must master, regardless of a vehicle's powertrain. But such equations are not the biggest challenges Farah and his team faced in developing the Volt.

"Probably the biggest started with the 'let's get all this stuff packaged in a vehicle this size,'" Farah says, and not only do you have to get it in, but it has to be crashworthy. "We're shooting for five-star crashworthiness all the way around—all twenty-five stars. So far everything says we can make it.

"I'm very happy with the results we've had. We've had a few glitches. We've gone back and made a little change and tweak, and we've been able to recover."

The other big engineering challenge from Farah's perspective was the necessity to make changes to systems that in other vehicles would be considered well and fully developed.

For example the interaction between the drivetrain and HVAC (heating, ventilation, and air conditioning) system, which, he says, in the case of the Volt is "fraught with significant invention."

The Chevrolet Volt takes to the streets of New York City for media test drives in conjunction with the 2010 New York International Auto Show. *GM*

The Chevrolet Volt visits San Francisco and cruises quietly down historic Lombard Avenue. *GM*

He explains, "In the typical car, the engine does it all. It provides heat and it turns the compressor. Here, we have an electric heater and an electric compressor. That's a very big change in how it gets done. And you're integrating that into a vehicle with all sorts of other issues. The whole thermal management aspect of the vehicle is extremely challenging."

And in more ways than you might expect. As an example, Farah shares details of one particularly frustrating challenge: Because there's not an engine that's running constantly, engineers had to develop an electric heating system for the Volt cabin. To avoid the inherent safety issues with electric heating systems that, for example, use an electrical coil to heat air (think of plug-in space heaters some people use in homes or garages), the Volt team opted for equipment that heats a liquid as it passes through a core and thus more closely resembles the radiator-based system used in the typical automobile. Farah says General Motors engineers developed what he called a "neat device" and applied for a number of patents.

"When it works, it does just what it should do," he says. The problem was that there were times when it didn't work as designed, and finding a solution became a big issue that finally reached crisis proportions.

Opposite: The Chevrolet Volt visits the Olympic Winter Games at Vancouver, British Columbia. *GM*

Basically, the system would seem to get itself into a nonoperating mode, and no one could figure out why.

Farah met with the team working on the issue and asked what they needed as they pursued a solution: More cars to test? More coffee to drink?

They checked and rechecked everything, down to the solder joints on the power drive integrated circuit that controlled the system. Finally, they looked again at the application notes that came with the integrated circuit and started dissecting the exact wording, which, Farah notes, "was probably translated from some other language." That's when they discovered that the instructions were not really very clear and may have led to the circuit board being installed incorrectly.

"We swapped a couple of wires around the other way, and it worked flawlessly," says Farah, who immediately ordered a new batch of boards to install in IVERs for validation. "With some things, it can be as simple as that. We moved on, and we haven't even talked about that part since then."

In addition to such hardware—or in this case difficult instructions issues—there are more esoteric issues to consider. Take, for example, passenger comfort.

"There's a huge range of how people like to feel," Farah says. And, in the case of the Volt, "you have to make the batteries feel comfortable on top of it."

But the Volt can carry only so much heating and cooling equipment.

"Who," he asks, "gets precedence, the battery or the customer?"

As it turns out, both do.

"The first buyers won't be nearly as concerned about it as the mainstream buyers who are coming," Farah says, knowing that even though the early adopters may be more flexible regarding staying warm in winter and cool in summer, "you have to make sure that the first ones don't have any weirdities, oddities, or 'interesting' service experiences."

Just as the car itself is more complicated, so is the engineering team.

"The makeup of the team is different because we're doing things you don't have to do otherwise," Farah says. "I have a battery and I have to know how much energy is in it, and there's a whole bunch of guys specializing in that. They are not the same guys who are working on our fuel tank or how to measure that. But those guys are on our team just like they do every fuel tank, but their methods and processes do not apply whatsoever to a battery. So there's a new set of people who have to be on the team. But all the other guys are still there, so the team is expanded greatly.

"There are more systems and more people to go with them, a lot more. . . . We're not making insignificant change here. We're making big significant change here."

And yes, he adds, "our goal is to make it just like any car."

So, have they succeeded? Is the Chevrolet Volt with all of its innovative technology really just like any car?

The only way to find out is to drive one, which is what I did early one weekend morning some 30 weeks before the start of production.

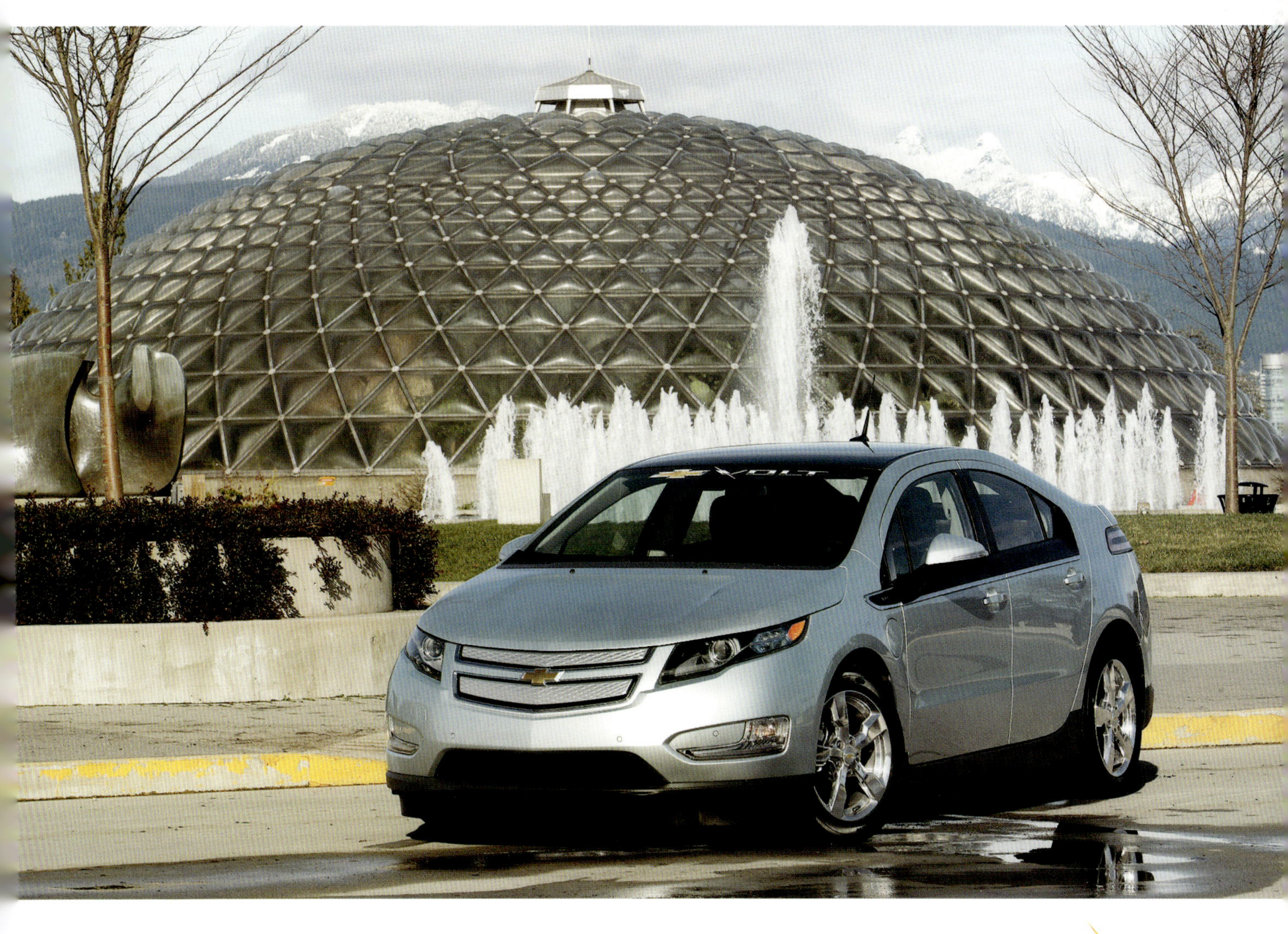

ANDREW FARAH AND HIS ENGINEERING TEAM *had spent much of the week before my test drive doing cold-weather testing in northern Ontario and doing HVAC validation work on the long drive back and forth from the Milford Proving Grounds. Farah says he felt sorry for those riding along with him because he used the 780-mile drive back from northern Ontario to push the extremes of personal discomfort so he was sure Volt owners won't have to.*

Farah allowed the cabin to get cold so that he could determine whether airflow within the cabin was balanced enough that both feet felt frozen at the same time. He also wanted to see, in real-world conditions, how using the HVAC system to maintain the most minimal occupant comfort might enhance the car's electric range.

"Every energy consumption opportunity affects range and fuel economy," says the chief engineer, who wants the production Volt to post a government-certified highway mileage figure that's closer to 50 miles per gallon than the anticipated 40-mile-per-gallon figure.

For those interested in such "hypermiling," the Volt offers HVAC controls that "the customer can understand and operate without having to be a PhD in energy balance equations," Farah says.

Again, he says, what GM learned from the EV1 is being applied to the Volt.

"In the EV1, you actually told the [HVAC] system how much power it could consume," Farah says. "There was a selector that had a middle position that said you can use no power to heat or cool, and it had about four levels on each side. It was a manual system. It was very good from the ability to control power but lousy from the ability to control comfort.

"On this car, we've gone the other way, a fully automatic system. Set it and forget it. But then we said, 'What about the guy who is willing to adjust personal comfort [as a trade-off for putting more energy into moving the vehicle]?' The answer is yes, we'll give them a power control. We actually gave them two additional levels—what we call comfort mode, where you can use the power to heat or cool the interior, and ECCO mode if you're willing to wear a coat [in winter] or sweat a little [in summer]," while using only half of the normal energy for interior heating and cooling. The cabin HVAC also can be turned off with only a fan bringing in unheated or uncooled exterior air.

"We've given the ability to have zero or to have half or to have full," Farah says.

Note: When the car comes from the factory, the comfort setting is engaged. "If they never touch it, that's what they'll use," Farah says.

The test drive began at 6:30 a.m. behind the wheel of one of the Volt IVER prototypes that had just returned from cold-weather testing in northern Ontario.

Farah and I met at a restaurant on the famed Woodward Avenue in one of Detroit's northwest suburbs, but it was so early the restaurant wasn't open yet, so we drove a couple of blocks and found another that was serving breakfast. We parked the Volt and went inside to talk, eat, and watch out the window as the car captured the attention of others who were up and out for an early meal.

Breakfast and interview over, we got into the car. Almost immediately, Farah pointed out that there was hot air coming through the vents even though the generator wasn't running.

The outside temperature indicator showed 36 degrees, so Farah switched the HVAC to its ECCO mode to demonstrate how the fan speed slows but the interior still stays very comfortable.

Putting the car into reverse triggers a backup camera.

"You could argue you don't need it," Farah says. "Mostly, it's the seats that block your view more than anything else."

There is a small lower rear window at the back of the car that enhances rearward visibility immediately behind the Volt, but Farah says experience shows that most people still look primarily through the upper window.

We shift into forward gear, pull out of the restaurant, and turn south toward the freeway.

In this IVER, we have to be careful as we move the shift control toward the dashboard because the car still is equipped with the original shifter design, which is beautiful and sort of mimics the power controller lever on a passenger jetliner. But while the device is beautiful, it also proved impractical because it didn't leave enough clearance for the driver's fingers when shifted into the Park position, so a redesign was done and already was being validated in several of the IVERs.

The display screen that replaces the gauge cluster in front of the driver shows various information, including speed and range, and a float-style indicator that encourages energy-efficient driving that involves more than just being gentle with the "gas" pedal. Accelerate too aggressively, and the float moves up out of the efficiency zone. Likewise, apply the brakes with too much force, and the float drops down below the efficiency zone.

Why? Because regenerative braking is a system that takes the power generated by braking but typically wasted in a conventional car and converts it to electricity that can be used to boost the charge in an electric- or hybrid-car's battery pack.

Turning heads: Tino Vasquez stops to take a look as Chris Sabatino takes a photo of the Chevrolet Volt at the historic Bob's Big Boy drive-in. *GM*

The Chevrolet Volt goes on display at Mann's Chinese Theater in Los Angeles, where Captain America, Batman, and the Flash consider its earth-friendly powertrain. *GM*

However, as Farah explains, "Brake too hard, and you're unable to recover all that energy."

We enter the freeway, zip down the ramp, and are moving at—even slightly above—the speed limit as quickly as we'd reach such rates in any sedan. With no engine noise, I'm slightly more aware of wind and tire noise. Basically, however, the Volt, even in this prototype stage that doesn't have factory fit and finish (or sound insulation), is luxury car quiet.

What seems most remarkable about the Volt is how unremarkable the driving experience is. It does not feel like a big golf cart. Even if it doesn't quite sound like a typical compact sedan, it feels like one. Steering response and feedback, braking, suspension reaction—all feel just like any other car.

A couple of days before my drive, Gery Kissel, the General Motors engineer in charge of developing the recharging hardware for the Volt, talked about his experience driving an EV1: "Electric vehicle propulsion is so much fun to drive. You fall in love it with as soon as you drive it. You'll have trouble going back. It's so much better. The driving sensation is so smooth and responsive, and to have that responsiveness from the instant torque!"

Accelerate from a stop in the Volt, and there's no waiting for an engine to rev up to its power curve. With electric motors, the power curve is more of an on/off switch. It's just always there when the motor is on.

The noticeable differences between the Volt and other compact sedans are experienced not so much by the seat of your pants but by your senses of sight and touch. They involve finishes like the electric-pattern design detail of the interior door panels, the presence of not one but two television-style display screens on the dashboard, the color and materials used for the instrumental panel, the touch-sensitive switchgear, and the way the panel cover curves inward from both sides toward a pair of small bubbles that pop out of the top of the dash near the bottom of the windshield.

Farah explains the second bubble on the dash, noting that General Motors vehicles typically have one such bubble that houses an ambient light sensor used by automatic HVAC systems. He explains that this bubble on the Volt is slightly taller because of the more dramatic rake of the windshield and because of the way the A-pillar shadows fall across the dashboard. The second bubble houses an indicator light that illuminates to provide a visible indicator when the car is plugged in and being charged.

Speaking of battery charge, the battery range indicator counts down as we drive along. We listen but really don't hear when the four-cylinder, internal combustion engine generator starts running, though the driver display changes to show that the generator has kicked in and the display now indicates a more traditional distance-to-empty range.

"We want it to be an unremarkable event [when generator comes on]," Farah says, proud of the fact that when the event happened, "you felt nothing. You heard nothing."

We work our way back toward town and continue the drive through a typical urban setting of residential neighborhoods and business-lined city streets.

At breakfast, Farah had talked about some of the Volt's unusual features, which include a power outlet built into the covered storage bin on top of the dashboard, an umbrella holder that's built into the driver's interior door panel, and a pedestrian-friendly horn.

"There is concern that when the vehicle is in electric mode, it is much quieter [than a typical car]," Farah says. "Say you're in one of those big-box store parking lots where people get out of their cars and walk down the middle of row toward the store. You're driving up behind them in a Volt, and you're not making any noise, and they don't know you're back there. So what do you do?

"You could blow the regular horn and scare the beejeebers out of them," he acknowledges, though that might lead to unpleasant reactions.

"You need a horn that's not a 'hey you!' but is more of an 'excuse me' type of horn, so we've made a horn that's much more friendly."

Instead of depressing the usual horn section of the steering wheel, what the Volt driver does is use the turn signal lever and its flash-to-pass feature that normally triggers a momentary flash of your bright lights. In the Volt, if you're driving at less than 25 miles per hour, you also get a quick, half-volume "chirp-chirp" from the horn.

Yes, Farah says, the Volt team considered using other sounds, but "you don't want it to sound like a piano or a bird chirping; you want it to sound like a car, just not as startling as someone just leaning on the horn."

The pedestrian-friendly horn is part of the Volt's safety protocol. So is the fact that when you open the hood, the engine generator starts up.

The engine generator starting when the hood opens is not new, Farah says, but is a feature also found on General Motors' hybrids. The goal is to prevent someone from opening the hood, reaching in, say, to check the oil or to do some other routine maintenance or even cleaning, and having an arm or hand in the way when the engine kicks on because the battery needs recharging. If you need to work under the hood, you can shut the engine generator off from inside the cockpit and not have to worry about it coming back on when you return to your chores.

Such a precaution might have been unnecessary, but it shows the concern the engineering team had to have to make the Volt ownership experience as pleasant as possible.

Here are two more examples:

There are people who may be able to drive a Volt without engaging the engine generator at all. Again, in Farah's mind this is another lunatic, fringe situation, nonetheless one that must be addressed. So the question is: How do you keep the engine generator properly lubricated and the gasoline in the tank from spoiling if the engine generator isn't used for an extended period?

"If you were to store a vehicle for a long time, everybody knows you ought to start it and run it up to temperature for a number of reasons to keep it lubricated, to force out any water that may be condensing in the crankcase, those kinds of things," Farah says. "That's condition number one.

Motor Trend magazine selected the Chevrolet Volt its 2011 car of the year. *GM*

But, in fact, that's precisely why the Chevrolet Volt is so remarkable, because driving it is so unremarkable.

"Condition number two is the concern that fuel will become stale.

"These days, with the quality of fuels that we have and the quality of the delivery system, that's not as much of an issue as it used to be. But the issue comes with what we call seasonal blending. A lot of places change the blend from summer to winter.

"So there if you think about it, you have about a six-month clock on the fuel. And on the engine, you probably have a roughly equivalent clock, maybe not quite so long if you really want to be a purist.

"So the question is: How often will the engine start? That's the question you have to ask yourself. For a very, very limited number of people, the engine will not start ever. These would be people who have the Volt as a second car or a third car, and they only drive it to local things and very purposefully. But there are very few people who will actually do that. The idea with the Volt that makes it different is that you don't have to worry about a lot of things, so you use it like you would use any old car. And most people, on a weekend, will typically go more than forty miles. I can stop one hundred people in a shopping mall, and I'll find two who don't, and that's fine. But we're not building this vehicle to be a specialty thing. We're building it so it can be your only car. That's why this discussion doesn't worry me. This is an antiseptic, academic discussion. So I don't worry about it.

"But the answer is, yes, we have two mechanisms in there. One will be a mechanism that will be watching the fuel age and the oil age, and the other will be watching the time elapsed since the engine ran . . . We will tune both those systems to (a) be unobtrusive and (b) do their job for those two guys out of a hundred.

"These are interesting discussions, but it's like talking about what tread should a shoe have on the moon. We send a couple of guys to the moon, and there's a tread on their shoe, and it has to work, but it really doesn't much matter what the pattern is."

Speaking of tread, the Volt rides on a unique Goodyear tire that enhances range through low rolling resistance, yet it stunned third-party engineers for its inclement weather grip when the car's slick-surface dynamics were being tested.

Farah notes that the tire is another area where the Volt is setting new industry standards.

"We've created a very efficient tire," he says of his team's work with Goodyear. "Of course, our tire is more expensive. We're paying thirty percent more for this tire, but it does so much. You get what you pay for."

That also can be said of the Volt itself. The car is not inexpensive, but it does so much, and so efficiently.

When our drive was finished, the car indicated that while the engine generator was supplying energy, that energy was moving us at an average of 52.9 miles per gallon.

Farah was pleased—his personal target for the car is 50 or better—but he also wondered if those who write about the Volt will really understand just what the car is all about.

"When people write about this car in the future," he says, "they'll write more about the handling and the NVH [noise, vibration, and harshness control] and those kind of things than they'll ever write about what's going on with the squirrels in the cages. We'll be boiled down to zero to sixty, to stopping distances, to decibels at road load, those kinds of things.

"It's just kind of funny, but it shows that all electronics is just a means to an end."

But, in fact, that's precisely why the Chevrolet Volt is so remarkable, because driving it is so unremarkable. It is not a different experience. It is simply a means to an end, a way to get from Point A to Point B, safely, comfortably, enjoyably, but without needing so much, if any, imported petroleum. It is, after all, a car, just as it was meant to be.

A car, of course, with a revolutionary propulsion system, but it can be driven just like any other.

A car that drives for miles and miles on electricity from the grid but that can just keep going after that electricity is consumed.

A car that just keeps on going, and going, and . . .

ACKNOWLEDGMENTS

The author is grateful that Bob Lutz, Jon Lauckner, Bob Boniface, Tim Greig, Jelani Aliyu, Ed Welburn, Stuart Norris, Frank Weber, Tony Posawatz, Andrew Farah, Nick Zielinski, Rich Lannen, Larry Laws, Nina Tortosa, Bob Kruse, Paul Pebbles, James Kobus, Gery Kissel, Teri Quigley, Nancy Laubenthal, Cristi Landy, Brent Dewar, Doug Parks, Pam Fletcher, Larry Kwiecinski, and Brian McGee of General Motors each took time to be interviewed for this book.

As always, Lutz was candid. Lauckner was a wealth of detail and knowledge. Farah was my hero, a chief engineer who could explain technicalities so even a journalist might understand. The same goes for Patil Prabhakar of Compact Power Inc. and his explanation of and insight into lithium-ion battery technology.

With young designers like Aliyu and engineers like Tortosa, I'm confident not only in projects like the Volt but in General Motors' future.

Speaking of confidence, those who buy the Volt will be grateful their cars were put together by Quigley and her team at the Detroit-Hamtramck assembly plant.

The author also acknowledges various public relations staffers who set up those interviews, including (and I apologize because I'm sure I'm missing someone) David Darovitz, Rob Peterson, Chris Preuss, Julie Houston-Rough, Robyn Henderson, Chris Lee, Brian Corbett, Jim Kobus, Kelly Wysocki, and Richard Pacini.

Steve Fecht and other photographers employed by General Motors took the photographs in the book. Beyond that, Fecht was a source of encouragement and help above and beyond. The same can be said of Paula Disalvo of GM Licensing and Lesley Tuttle of Equity Management.

The author also thanks publisher Zack Miller and editor Chris Endres of Motorbooks, and acknowledges Endres' remarkable patience while waiting for the manuscript to be completed.

Index